SPRINGER PROTOCOLS HANDBOOKS

For further volumes:
http://www.springer.com/series/8623

Springer Protocols Handbooks collects a diverse range of step-by-step laboratory methods and protocols from across the life and biomedical sciences. Each protocol is provided in the Springer Protocol format: readily-reproducible in a step-by-step fashion. Each protocol opens with an introductory overview, a list of the materials and reagents needed to complete the experiment, and is followed by a detailed procedure supported by a helpful notes section offering tips and tricks of the trade as well as troubleshooting advice. With a focus on large comprehensive protocol collections and an international authorship, Springer Protocols Handbooks are a valuable addition to the laboratory.

Endosymbionts

Edited by

N. Amaresan and Sapna Chandwani

C.G. Bhakta Institute of Biotechnology, Uka Tarsadia University, Surat, Gujarat, India

Humana Press

Editors
N. Amaresan
C.G. Bhakta Institute of Biotechnology
Uka Tarsadia University
Surat, Gujarat, India

Sapna Chandwani
C.G. Bhakta Institute of Biotechnology
Uka Tarsadia University
Surat, Gujarat, India

ISSN 1949-2448 ISSN 1949-2456 (electronic)
Springer Protocols Handbooks
ISBN 978-1-0716-4770-7 ISBN 978-1-0716-4771-4 (eBook)
https://doi.org/10.1007/978-1-0716-4771-4

This Humana imprint is published by the registered company Springer Science+Business Media, LLC, part of Springer Nature.
The registered company address is: 1 New York Plaza, New York, NY 10004, U.S.A.

Preface

Endosymbionts, the organisms that reside within the cells or tissues of other host species, play essential roles in shaping biological systems, from aiding host metabolism to influencing evolution and ecological balance. Their presence supports various life forms, including fungi, algae, bacteria, and insects, making the study of endosymbionts complex and multidisciplinary. As research into endosymbionts is in progress, there is an increasing demand for practical methodologies and protocols to explore and harness their potential.

This book is composed of 4 sections with 28 protocol chapters. Section I focuses on the isolation and maintenance of diverse endosymbionts and offers protocols for working with fungi, algae, actinobacteria, mycorrhizae, lichens, and insect hosts. Section II presents a variety of identification techniques, ranging from classical microscopy and staining to advanced molecular and single-cell approaches such as PCR, FISH, flow cytometry, and next-generation sequencing. Section III explores the biological potential of endosymbionts by outlining methods to screen for metabolites with applications in agriculture, medicine, and biotechnology, including plant growth promotion, antimicrobial activity, and biocontrol. Section IV details specialized methods for detecting bacterial endosymbionts in insects, which are crucial for understanding host-pathogen interactions and symbiotic balance. This handbook is designed as a hands-on guide for undergraduate and postgraduate students, researchers, and practitioners working in microbiology, symbiosis, biotechnology, and related fields. Each chapter is contributed by experienced researchers who provided step-by-step procedures, practical tips, and insights gained from years of experimentation. The handbook not only serves as a manual but also encourages the exploration of endosymbionts as a source of novel bioactive compounds and ecological insights.

The editors thank all contributors for their valuable input. We hope that this book will empower researchers to expand the frontiers of endosymbiosis research and unlock new possibilities for science and society.

Surat, Gujarat, India

N. Amaresan
Sapna Chandwani

Contents

Part IV Methods to Detect Bacterial Infection of Endosymbionts in Insects

Contributors

NATARAJAN AMARESAN • *C.G. Bhakta Institute of Biotechnology, Uka Tarsadia University, Bardoli, Surat, Gujarat, India*

T. S. ARCHANA • *Department of Plant Pathology, School of Agriculture, Lovely Professional University, Phagwara, Punjab, India*

KOMAL CHANDARANA • *C.G. Bhakta Institute of Biotechnology, Uka Tarsadia University, Bardoli, Surat, Gujarat, India*

K. S. V. P. CHANDRIKA • *Division of Crop Protection, ICAR-Indian Institute of Oilseeds Research (ICAR-IIOR), Hyderabad, India*

SAPNA CHANDWANI • *C.G. Bhakta Institute of Biotechnology, Uka Tarsadia University, Bardoli, Surat, Gujarat, India*

DIRAYAH RAUF HUSAIN • *Department of Biology, Faculty of Mathematics and Natural Sciences, Universitas Hasanuddin, South Sulawesi, Indonesia*

SAMIKSHA JHAMTA • *Department of Zoology, Akal College of Basic Sciences, Eternal University, Baru Sahib, Himachal Pradesh, India*

SIMRANJEET KAUR • *Department of Zoology, Akal College of Basic Sciences, Eternal University, Baru Sahib, Himachal Pradesh, India*

SUNIL H. KOLI • *Department of Microbiology, Yashwantrao Chavan College of Science, Karad, MH, India*

DEVENDRA KUMAR • *Department of Plant Pathology, School of Agriculture, Lovely Professional University, Phagwara, Punjab, India*

GAGAN KUMAR • *Department of Plant Pathology, Acharya Narendra Deva University of Agriculture and Technology, Kumarganj, Ayodhya, Uttar Pradesh, India*

SARIKA KUMARI • *C.G. Bhakta Institute of Biotechnology, Uka Tarsadia University, Bardoli, Surat, Gujarat, India*

GANDRAJAPALLY SHIVAREDDY MADHU • *CHES-Chettalli, ICAR-Indian Institute of Horticultural Research, Bengaluru, Karnataka, India*

SURESH MADUGULA • *Division of Crop Protection, ICAR-Indian Institute of Oilseeds Research (ICAR-IIOR), Hyderabad, India*

SEPHOKOANE C. MAKUWA • *Faculty of Science, Department of Biotechnology and Food Technology, University of Johannesburg, Johannesburg, South Africa*

KOTIPALLI SATYA NAGA MANIDEEP • *Division of Crop Protection, ICAR-Indian Institute of Horticultural Research, Bengaluru, Karnataka, India*

LAKSHMAIAH MANJUNATHA • *Division of Crop Protection, ICAR-Indian Institute of Horticultural Research, Bengaluru, Karnataka, India*

PFARISO MAUMELA • *Faculty of Science, Department of Biotechnology and Food Technology, University of Johannesburg, Johannesburg, South Africa*

RAJVI MEHTA • *C.G. Bhakta Institute of Biotechnology, Uka Tarsadia University, Surat, Gujarat, India*

BHAVANA V. MOHITE • *Department of Microbiology, Bajaj College of Science, Wardha, MH, India*

LIZZY A. MWAMBURI • *University of Eldoret, Eldoret, Kenya*

HETVI NAIK • *C.G. Bhakta Institute of Biotechnology, Uka Tarsadia University, Bardoli, Surat, Gujarat, India*

JINAL NAIK • *Department of Biotechnology, School of Science and Technology, Vanita Vishram women's University, Surat, Gujarat, India*

PRADNYA B. NIKAM • *School of Life Sciences, KBC North Maharashtra University, Jalgaon, MH, India*

ATHARVA S. PATIL • *Department of Pharmacy, H R Patel College, Shirpur, Maharashtra, India*

SATISH V. PATIL • *School of Life Sciences, KBC North Maharashtra University, Jalgaon, Maharashtra, India*

VIKAS S. PATIL • *University Department of Chemical Technoogy, KBC North Maharashtra University, Jalgaon, Maharashtra, India*

R. D. PRASAD • *Division of Crop Protection, ICAR-Indian Institute of Oilseeds Research (ICAR-IIOR), Hyderabad, India*

K. V. BHASKARA RAO • *Marine Biotechnology Laboratory, Department of Bio-Medical Sciences, School of Bio Sciences and Technology, Vellore Institute of Technology, Vellore, Tamil Nadu, India*

SHAIK RESHMA • *Division of Crop Protection, ICAR-Indian Institute of Oilseeds Research (ICAR-IIOR), Hyderabad, India*

SHRADDHA SAHA • *The Mandvi Education Science College, Veer Narmad South Gujarat University, Surat, Gujarat, India*

JITENDRA D. SALUNKHE • *School of Life Sciences, KBC North Maharashtra University, Jalgaon, Maharashtra, India*

NARENDRA S. SALUNKHE • *School of Life Sciences, KBC North Maharashtra University, Jalgaon, MH, India*

SHIMPY SARKAR • *Department of Entomology, School of Agriculture, Lovely Professional University, Phagwara, Punjab, India*

MAHLORO H. SEREPA-DLAMINI • *Faculty of Science, Department of Biotechnology and Food Technology, University of Johannesburg, Johannesburg, South Africa*

ARSHDEEP SINGH • *Department of Agronomy, School of Agriculture, Lovely Professional University, Phagwara, Punjab, India*

SHIVAM SINGH • *Krishi Vigyan Kendra-Baghpat, S.V.P University of Agriculture and Technology, Meerut, Uttar Pradesh, India*

NEELAM THAKUR • *Department of Zoology, Akal College of Basic Sciences, Eternal University, Baru Sahib, Himachal Pradesh, India*

P. V. THARANI • *Marine Biotechnology Laboratory, Department of Bio-Medical Sciences, School of Bio Sciences and Technology, Vellore Institute of Technology, Vellore, Tamil Nadu, India*

KURACHA TULASI PRIYA • *Division of Crop Protection, ICAR-Indian Institute of Horticultural Research, Bengaluru, Karnataka, India*

VENKATARAVANAPPA VENKATARAVANAPPA • *Division of Crop Protection, ICAR-Indian Institute of Horticultural Research, Bengaluru, Karnataka, India*

RIUH WARDHANI • *Department of Biology, Faculty of Mathematics and Natural Sciences, Universitas Hasanuddin, South Sulawesi, Indonesia; Department of Animal and Dairy Science, College of Agriculture and Life Sciences, Chungnam National University, Daejeon, Republic of Korea*

KAMLESH K. YADAV • *Mata Jijabai Government P G Girls College, Indore, Madhya Pradesh, India; Green Grahi Solution Pvt. Ltd., Roorkee, Uttarakhand, India*

PRATIBHA YADAV • *Mata Jijabai Government P G Girls College, Indore, Madhya Pradesh, India; Green Grahi Solution Pvt. Ltd., Roorkee, Uttarakhand, India*

Part I

Isolation Methods of Endosymbionts

Chapter 1

Isolation and Maintenance of Endohyphal Symbionts from the Fungi

Suresh Madugula, Shaik Reshma, K. S. V. P. Chandrika, and R. D. Prasad

Abstract

Fungi are both widespread and essential members of environmental and host-associated microbiomes. Their presence and functional roles have a significant impact on microbiome dynamics and broader ecological processes. Despite this, the biological complexity of fungi is often underestimated. Fungi generate a variety of intricate structures throughout their life cycle and contain numerous organelles, lipid droplets, and other intracellular components. Additionally, filamentous fungi can host bacteria (such as cyanobacteria), mycoviruses, and other fungi and are even capable of internalizing entire microalgal cells within their hyphae.

Endohyphal symbionts are microorganisms, such as bacteria, that live inside the hyphae of fungi. Isolation and maintenance of these symbionts require careful handling to ensure that the symbionts are separated from the host fungi and maintained under appropriate conditions. In this book chapter, we describe the various protocols for isolating and maintaining endohyphal symbionts from fungi.

Key words Selective media, Symbiotic organisms, Bacterial endophytes, Isolation, Staining

1 Introduction

Throughout their lifetime, plants engage with a vast array of microorganisms, including archaea, bacteria, fungi, and viruses. However, only a few selective microorganisms form symbiotic relationships with their plant hosts. Fungi, in particular, can establish symbiotic associations with plants, influencing the host's fitness in either beneficial or detrimental ways. In addition to the association between plants and microorganisms, symbiotic relationships also occur between bacteria and plant-associated fungi. Studies on these bacterial–fungal symbioses have identified two categories of bacterial symbionts: bacterial ectosymbionts and endosymbionts. These symbionts differ based on their physical interactions with the host fungus. Bacterial endosymbionts, also referred to as endofungal or endohyphal bacteria, represent the most intimate form of symbiosis between bacteria and fungi [1–3].

N. Amaresan and Sapna Chandwani (eds.), *Endosymbionts*, Springer Protocols Handbooks,
https://doi.org/10.1007/978-1-0716-4771-4_1,

Endohyphal symbionts are microorganisms, typically bacteria, that reside within the hyphae (filamentous structures) of fungi. This symbiotic relationship can influence the biology, ecology, and evolution of both the fungi and the associated microorganisms [4]. These symbionts can affect the host fungus's growth, pathogenicity, stress tolerance, and metabolic capabilities. They may also influence the interactions of the fungus with its environment, including plant hosts, other microorganisms, and the soil ecosystem.

1.1 Types of Symbiotic Relationships

Mutualism: In many cases, endohyphal symbionts provide benefits to their fungal hosts, such as enhancing nutrient acquisition, stress tolerance, or resistance to pathogens (*see* Fig. 1).

Commensalism: Some endohyphal symbionts may reside within the fungal hyphae without causing any apparent benefit or harm (*see* Fig. 1).

Parasitism: In some instances, the relationship may be detrimental to the fungus, with the symbiont exploiting the fungal resources (*see* Fig. 1).

The most challenging task is the isolation and culturing of certain endohyphal symbionts, as they may have specific growth requirements or may be obligate symbionts that do not grow outside the fungal host. Isolating endohyphal symbionts in agriculture involves extracting microorganisms, such as bacteria, that live

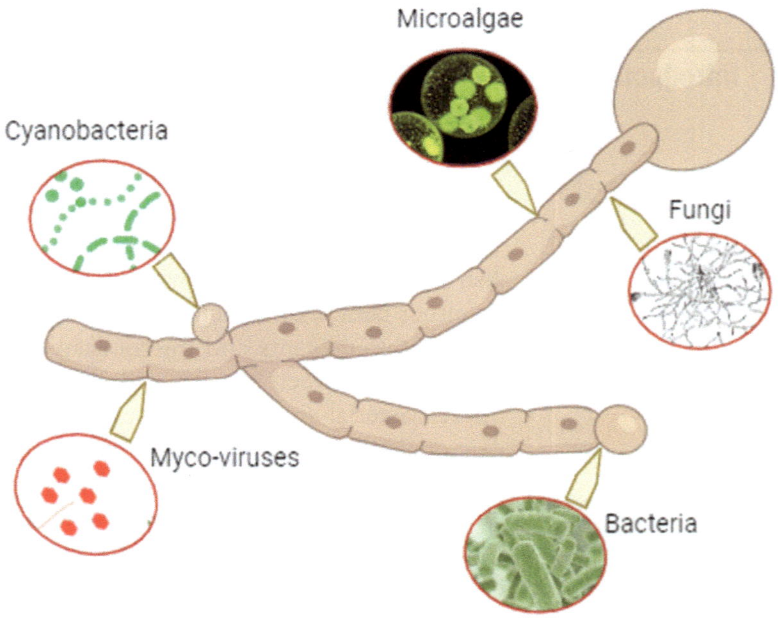

Fig. 1 The most common endohyphal symbionts—microalgae, mycoviruses, bacteria (including cyanobacteria), and fungi. (Created in Biorender.com)

within the hyphae of fungi, particularly those associated with plants. This process is critical for studying the symbionts' role in plant health and developing biocontrol agents or biofertilizers.

Moreover, *selective media* are crucial for isolating endohyphal symbionts because they promote the growth of target microorganisms while inhibiting unwanted ones.

2 Materials

2.1 Surface Sterilization of Fungal Material

- Ethanol (70%).
- Sodium hypochlorite solution (1–3%).
- Sterile distilled water.
- Sterile forceps, scalpel, and Petri dishes.

2.2 Isolation of Hyphae and Symbiont Release

- Sterile Petri dishes or slides.
- Sterile dissecting tools (scalpels, forceps, needles).
- Dissecting microscope.

2.3 Isolation and Purification of Endohyphal Symbionts

- Autoclave.
- Sterile selective media agar plates (e.g., Luria broth agar, yeast extract mannitol agar).
- Sterile inoculating loops or needles (*see* **Note 1**).

2.4 Selective Media for Isolating Endohyphal Symbionts

2.4.1 Reasoner's 2A Agar (R2A Agar)

Add yeast extract 0.5 g; proteose peptone 0.5 g; casein hydrolysate 0.5 g; glucose 0.5 g; soluble starch 0.5 g; sodium pyruvate 0.3 g; dipotassium hydrogen phosphate 0.3 g; magnesium sulfate anhydrous 0.024 g; agar–agar 15.0 g in 1 L of water. Adjust the pH to 7.0 with 1 N NaOH and autoclave the mixture for 25 min at 120 °C [5–7].

2.4.2 Luria–Bertani Agar (LB Agar)

(a) From individual components: (1) For each 950 mL of MilliQ H_2O, add 10 g tryptone, 10 g sodium chloride (NaCl), and 5 g yeast extract. (2) Mix until powder is dissolved. (3) pH solution to ~7.0 using sodium hydroxide (NaOH). (4) Adjust the solution to a final volume of 1000 mL. (5) Autoclave using liquid cycle. Add 25 g of a premixed formulation containing tryptone, NaCl, and yeast extract in 950 mL of Milli-Q H_2O, and mix it well until the powder is dissolved. Adjust the pH solution to ~7.0 using sodium hydroxide (NaOH) and make a final volume of 1000 mL. Autoclave the media and store it at room temperature for further use [8]. This medium is effective for the initial cultivation of general bacterial symbionts from fungal hyphae.

2.4.3 King's B Agar

Dissolve 33 g premixed powder in 990 mL distilled water and add 10 mL glycerol. Sterilize the media by autoclaving at 121 °C for 15 min [9, 10].

2.4.4 Actinomycete Isolation Agar (AIA)

Suspend 21.7–22 g of the premixed powder containing agar powder 15 g/L, L-asparagine 0.1 g/L, dipotassium phosphate 0.5 g/L, ferrous sulfate 0.001 g/L, magnesium sulfate 0.1 g/L, sodium caseinate 2 g/L, and sodium propionate 4 g/L in 1 L of distilled water. Add 5 mL of glycerol and boil the mixture to completely dissolve the powder. Autoclave at 121 °C for 15 min [11]. This medium is effective for isolating endohyphal actinomycetes involved in antibiotic production or other symbiotic functions.

2.4.5 Nitrogen-Free Agar (NFA)

1. *Ashby Nitrogen Free Agar Plate/Ashby's Mannitol Agar* (0.2 g KH_2PO_4, 5 g $CaCO_3$, 0.2 g $MgSO_4 \cdot 7H_2O$, 10 g mannitol, 0.2 g NaCl, 15 g agar, 0.1 g $CaSO_4 \cdot 2H_2O$ per liter, pH 7.0.)
2. *Dobereiner Nitrogen-Free Agar Plate* (5 g malic acid, 0.4 g $KH_2PO_4 \cdot H_2O$, 0.2 g $MgSO_4 \cdot 7H_2O$, 0.1 g NaCl, 0.02 g $CaCl_2 \cdot H_2O$, 0.01 g $FeCl_3$, 0.002 g $Na_2MoO_4 \cdot 2H_2O$ per liter; pH 7.0.)

This medium is considered for isolating endohyphal bacteria involved in nitrogen fixation, particularly relevant for symbionts of mycorrhizal fungi associated with plant roots.

2.4.6 Chitin Agar

Mix colloidal chitin (10.0 g), NH_4Cl (5 g), KH_2PO_4 (2.4 g), $MgSO_4.7H_2O$ (0.5 g), K_2HPO_4 (0.6 g), and bacteriological agar (15 g) in 1 L distilled water, autoclave at 121 °C for 15 min and cool to 50 °C. Adjust the pH to 7.5–9.0 with 5 N NaOH [12]. The medium is considered to be ideal for isolating endohyphal bacteria that may rely on or degrade fungal cell wall components.

2.4.7 Yeast Extract Mannitol Agar (YEM)

Weigh out 0.5 g of yeast extract and 15 g of agar powder. Add this mixture to 869 mL of Milli-Q H_2O. Autoclave the mixture at 120 ° C for 30 min [13]. The medium is useful for isolating endohyphal bacteria that might form symbioses similar to rhizobia, potentially aiding fungal or plant hosts.

2.4.8 Tryptic Soy Agar (TSA)

Suspend 45 g premixed powder (15 g pancreatic digest of casein; 5 g peptic digest of s*oybean*; 5 g sodium chloride; 15 g *agar*) in 1000 mL distilled water. Heat to boiling to dissolve the medium completely. Sterilize by autoclaving at 15 lbs. pressure (121 °C) for 15 min. Cool to 45–50 °C [14]. The medium is ideal for isolation when the specific nutritional requirements of the endohyphal symbionts are unknown.

2.4.9 Selective Antibiotic-Containing Media

General growth media (e.g., potato dextrose agar, malt extract agar, tryptic soy agar) supplemented with specific antibiotics (e.g., tetracycline, streptomycin, nystatin). The medium is effective for isolating bacteria when the fungal host or other bacteria are susceptible to the antibiotics used [15].

2.4.10 Low-Nutrient Agar (e.g., 1/10th Strength Tryptic Soy Agar)

The medium is beneficial for isolating endohyphal bacteria adapted to low-nutrient conditions within fungal hyphae.

2.4.11 Selective Media

The choice of selective media should consider the specific fungal host and its associated symbionts, as different fungi may harbor distinct types of bacteria or fungi with varying nutrient requirements (*see* **Note 2**).

2.4.12 Media Modifications: (See **Note 3**)

2.5 Verification of Endohyphal Symbionts

2.5.1 Microscopy

Compound microscope (100–400×) or phase-contrast or fluorescence microscope.

2.5.2 Florescence Imaging/Epifluorescence Microscopy

DAPI (4′,6-diamidino-2-phenylindole), phosphate buffer solution, 4% paraformaldehyde.

2.5.3 Flow Cytometry

Live/dead stain (calcein–AM, propidium iodide, and SYTO 9).

2.5.4 Molecular Identification

DNA extraction and PCR methods followed by sequencing (described in detail in methodology).

2.6 Antibiotic Sensitivity Testing

- Tryptic soy agar (TSA) and King's B (KB) medium agar plates.
- Antibiotic filter paper discs (each containing different antibiotics, viz. penicillin, cephradine, clindamycin, erythromycin, kanamycin, ampicillin, ciprofloxacin, gentamicin, doxycycline, amoxicillin, vancomycin, ceftriaxone, chloramphenicol, cotrimoxazole).

2.7 Maintenance of Isolated Endohyphal Symbionts

- Glycerol.
- Fridge (short term storage)/-20 °C deep freezer (long term storage).
- Lyophilizer.

3 Methods

3.1 Sample Collection

- Collect fungal samples from the environment using sterile tools to prevent contamination.
- Place the collected samples in sterile containers for transport to the laboratory.
- Harvest plant roots or other tissues that are known to harbor mycorrhizal fungi or endophytes. Make sure to collect these samples using sterile scissors or knives.
- Note the environmental conditions (e.g., soil type, moisture, temperature) as they can influence the fungal community and associated endohyphal symbionts.

3.2 Isolation and Purification of Endohyphal Symbionts

Isolation and purification involve several steps to ensure that the microorganisms residing within fungal hyphae are successfully separated from their host and other potential contaminants. The goal is to obtain pure cultures of these symbionts for further study and application. The following steps need to be followed for isolation and purification of endohyphal symbionts from plant samples.

3.2.1 Surface Sterilization of Plant Material with Endohyphal Symbionts

- To eliminate external contaminants, ensure that isolated symbionts are truly endohyphal (living within the fungal hyphae).
- Wash the healthy leaves or photosynthetic stems with microphylls in running tap water for 30 s (as detailed below) to remove epiphyllous debris, allow it to dry, and then cut into small fragments (231 mm).
- Rinse the plant material thoroughly in sterile distilled water (at least three to five times) to remove any residual bleach, which could affect both the fungus and its symbionts.
- Surface sterilize the tissue fragments using sequential immersion in 95% ethanol (10 s), 10% chlorine bleach (0.525% NaOCl; 2 min), and 70% ethanol (2 min), and allow it to surface-dry under sterile conditions (*see* **Note 4**).

3.2.2 Isolation of Hyphal Fragments or Spores from the Seeds or Plant Parts

- For isolation, cut the seed/plant part into half and allow it to surface-dry under sterile conditions.
- Place one-half of the seed on 2% malt extract agar (MEA) as per the protocol of Gallery et al. [16].
- Incubate the seed or the plant until the fungal isolate, containing its endohyphal symbiont, will grow out from the interior of the seed/plant part.
- Carefully cut small hyphal fragments (1–2 mm) under a dissecting microscope, or isolate spores from the sterilized fungal material using sterile tools.

- Place hyphal fragments or spores on a sterile surface, like a Petri dish lid or sterile filter paper, for subsequent transfer to growth media.

3.2.3 Inoculation on Selective Media

- Transfer hyphal fragments or spores to selective agar plates using sterile forceps or needles.
- Gently press the fungal material onto the agar surface to ensure good contact.
- Seal the plates with parafilm, or place them in airtight containers to prevent contamination.
- Incubate the plates at an appropriate temperature (usually 20–25 °C) for fungal growth and release of symbionts.
- Incubate the fungal hyphae on the medium at an appropriate temperature (usually around 20–25 °C) for 3–7 days to allow the growth of the fungal mycelium and release of endohyphal symbionts.
- Monitor on a daily basis to observe bacterial colonies emerging from the hyphal fragments, indicating the presence of endohyphal symbionts.

3.2.4 Purification of Endohyphal Symbionts

- To obtain pure cultures of endohyphal symbionts for further analysis, pick individual colonies that emerge from the hyphal fragments using a sterile inoculating loop or needle. These colonies likely represent endohyphal symbionts.
- Streak each colony onto a fresh agar plate to isolate pure cultures. Perform streaking in a zigzag pattern to spread out individual cells in order to get the isolated colonies.
- Incubate the streaked plates at the appropriate temperature (typically 28 °C for most bacteria).
- After 24–48 h, examine the plates for isolated colonies. If multiple types of colonies are present, repeat the streaking process with colonies that are morphologically similar and well separated to ensure purity.
- If there is any doubt about the purity of an isolated culture, perform re-streaking on fresh agar plates until a single, uniform colony type is obtained (*see* **Note 5**).

3.2.5 Verification of Endohyphal Symbionts

To confirm that the isolated microorganisms are true endohyphal symbionts and not contaminants, verification can be assessed via.

Microscopy

- Use a light microscope (100–400×)/phase-contrast or fluorescent microscope to observe bacteria within fungal hyphae directly. Staining methods like DAPI (4′,6-diamidino-2-phenylindole) can highlight bacterial DNA, confirming their location inside the hyphae.

Molecular Identification

- Extract DNA from the isolated bacterial cultures using a suitable DNA extraction kit/standard DNA extraction protocol [4, 17–20]. To extract total genomic DNA, remove small quantities of mycelium (75 μL in volume) directly from active cultures placed on 2% MEA, grind it with a sterile plastic pestle in 500 μL SDS extraction buffer (0.15 M NaCl, 50 mM Tris–HCl [pH 8.0], 10 mM Na_2EDTA, and 2% SDS), and incubate for 12 h at room temperature [21].
- Treat the samples with 500 μL of phenol:chloroform:isoamyl alcohol, mix it briefly, and centrifuge at 10,000 rpm for 15 min.
- Precipitate the DNA from the reserved supernatant using absolute isopropanol at −20 °C (0.6:1 isopropanol:aqueous volume), followed by centrifugation (5 min at 10,000 rpm), washing in 70% ethanol (−20 °C), drying for 40 min in a rotary vacufuge, and elution in 50 μL of sterile water.
- Dilute the extracted DNA in a ratio of 1:10 to perform PCR.
- Perform PCR amplification of the 16S rRNA gene (for bacteria) or ITS region (for fungi) using specific primers as detailed herewith:
  - Perform PCR amplification of the nuclear ribosomal internal transcribed spacer region (ITS rDNA) in 25 μL reaction volumes containing 2.5 μL dNTPs (10 μM), 2.5 μL bovine serum albumin (BSA), 2.5 μL PCR buffer, 1.25 μL of each primer (10 μM), 13.875 μL water, 0.125 μL Taq polymerase, and 1.0 μL of diluted template.
  - Use universal primers ITS5 or ITS1F and ITS4 or LR3 to amplify ITS rDNA.
  - Run the cyclic reactions on any standard thermal cycler using the following standardized parameters or after standardization as per lab conditions: 4 min at 95 °C; 35 cycles of 30 s at 95 °C, 30 s at 50°, 52°, or 54 °C, and 90 s at 72 °C; and 10 min at 72 °C.
  - Perform the gel electrophoresis on 1% agarose gel using SYBR green to visualize the obtained bands.
- Sequence the PCR products, compare the sequences to databases (e.g., NCBI BLAST) to identify the bacterial or fungal species and confirm their identity as endohyphal symbionts (described as hereunder):
  - Clean the PCR products using qiagen or microcon PCR purification columns and sequence them in 10 μL reaction volumes (1.5 μL of big dye, 2.5 μL of big dye buffer, 1 μL of primer, 3.5 μL of sterile water, and 1.5 μL of PCR product).

- Perform the sequencing of the amplified DNA for both forward and reverse reads on an automated sequencer using primers ITS5 or ITS1F and ITS4.
- Check all sequencing reads manually before submission to BLAST searches of the NCBI GenBank database for provisional identification at higher taxonomic levels.

Visual Confirmation through Staining Techniques

Florescence Imaging/ Epifluorescence Microscopy: DAPI Staining

- Fix the sample using 4% paraformaldehyde and permeabilize the cultured cells.
- Wash the cells with PBS for at least one to three times.
- Add 300 mM DAPI stain solution.
- To prepare a DAPI staining solution—a fluorophore—mix 5 μL of DAPI (10 mg/mL solution) with 995 μL of phosphate buffer (PB). This solution can be used for 1 month if stored at 4 °C in the dark. Add 10 μL of the solution to 990 μL of PB to make a working solution to cover the cells.
- Incubate the cells for 1–5 min in a place protected from light.
- Remove the stain solution.
- Again, wash the cells with PBS for at least two to three times.
- Image the cells (Fig. 2) [22].

Flow Cytometry: Live/Dead Staining

- Evaluate the fungal hyphae by using the live/dead BacLight bacterial viability kit (*Invitrogen*) as per the method of Hoffman and Arnold [2].
- To prepare fungal samples for visualization, scrap the fresh hyphae from the surface of the growing edge of each fungal colony on 2% MEA.
- Prepare axenic bacteria by scraping a single colony from LB agar.
- Place the hyphae or bacterial cells on a glass slide with 15 μL of 1:1:18 (components A and B along with sterile water in the kit) live/dead stain [23], then cover it with a coverslip, and incubate in darkness for 20 min.
- Remove excess water using a blotting paper.
- Seal the slides with clear acrylic nail polish, and allow them to dry before viewing.
- Use at least a 4000 MB compound microscope with a 100-W mercury arc lamp for fluorescent imaging at room temperature with a chroma technology 35,002 filter set (480-nm excitation/ 520-nm emission) and a 100× APO oil objective.

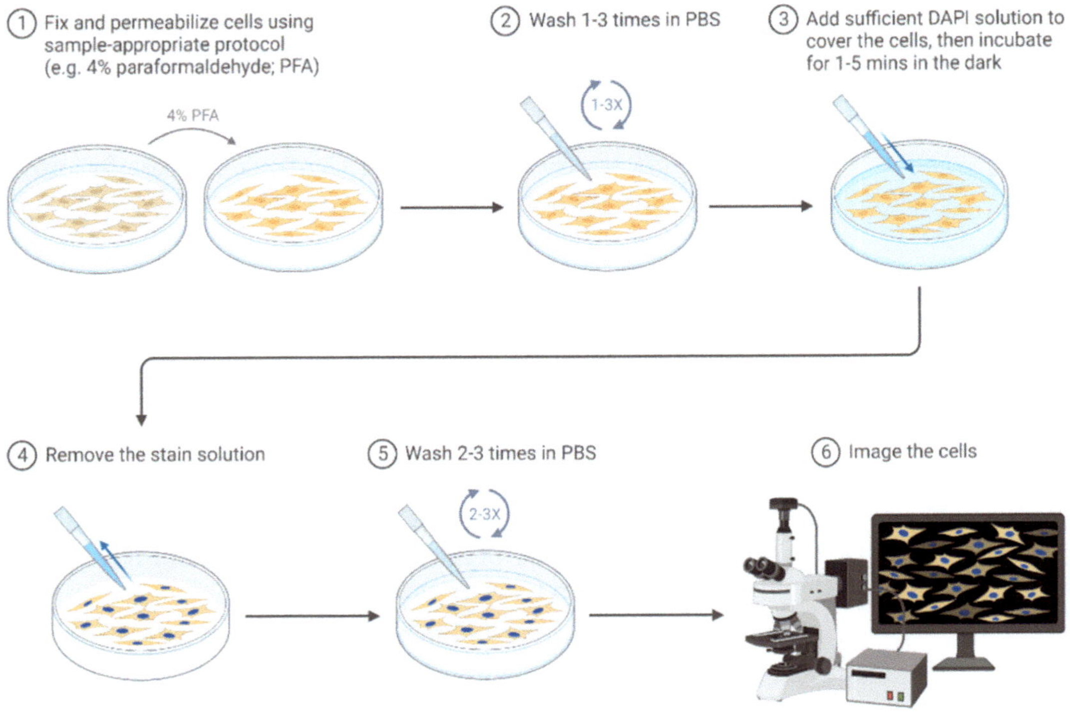

Fig. 2 Steps in DAPI staining. (Source: biorender.com)

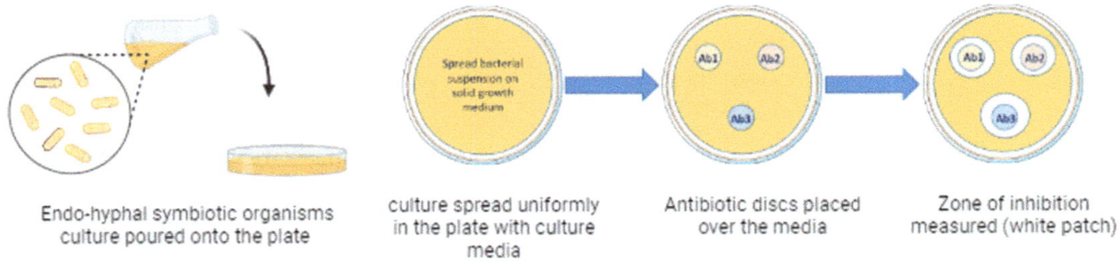

Fig. 3 Antibiotic sensitivity testing procedure. (Created in Biorender.com)

Antibiotic Sensitivity Testing

Test the isolated bacteria for antibiotic sensitivity to distinguish symbionts from potential contaminants. Symbionts often exhibit unique resistance profiles due to their association with fungal hosts. Antibiotic sensitivity tests for endophytes can be performed using the Kirby–Bauer disc diffusion method [24], which involves

- Use of antibiotic-impregnated discs.
- To test the strains against antibiotics like chloramphenicol, erythromycin, rifampicin, polymixin B, streptomycin, tetracycline, amoxicillin, and neomycin.
- To record the diameter of the inhibition zone (*see* **Note 6**).
- To categorize the organisms as resistant, intermediate, or sensitive based on the DIFCO manual (*see* Fig. 3).

3.3 Maintenance of Isolated Endohyphal Symbionts

3.3.1 Culturing

- Subculture the isolated bacteria regularly on suitable nutrient media. This can vary depending on the bacterial species but generally includes nutrient agar or LB agar.
- Incubate at an appropriate temperature (usually 28–37 °C, depending on the bacterial species).

3.3.2 Storage

- Keep bacterial plates or slants at 4 °C for short-term storage (1–2 months).
- For long-term storage, freeze bacterial cultures in 15% glycerol at −80 °C, or lyophilize (freeze-dry) them (*see* **Note 7**).

3.3.3 Verification of Endohyphal Origin

- To confirm that the isolated bacteria are indeed endohyphal symbionts, perform molecular identification using techniques like PCR amplification and bacterial 16S rRNA gene sequencing.

Use microscopy or staining methods to observe the presence of bacteria within the fungal hyphae before isolation.

4 Notes

1. Always use aseptic techniques to avoid contamination during the entire process.
2. The choice of media and incubation conditions should be tailored to the specific fungi and symbionts being studied, as different organisms have different growth requirements.
3. Some endohyphal symbionts may require specialized media modifications, such as adjusted pH levels, specific carbon or nitrogen sources, or reduced oxygen levels, to mimic the conditions within fungal hyphae.
4. For surface sterilization, the following steps can be followed: *Ethanol rinse:* Briefly immerse the plant material (e.g., mycelium or spores) in 70% ethanol for about 30 sec to 1 min. *Bleach treatment:* Submerge the plant material in a 1–3% sodium hypochlorite solution for 1–5 min. The exact duration depends on the toughness of the fungal tissue.
5. For bacteria or fungi that do not readily separate on streak plates, use a micromanipulator or a serial dilution and plating method to isolate single cells or spores. These can then be cultured individually to ensure purity.
6. Include negative controls (no fungal material) to ensure that any observed bacterial growth originates from the fungi and not from environmental contamination.

7. Follow all laboratory safety guidelines and protocols, especially when handling potentially pathogenic fungi or bacteria.

References

1. Bastías DA, Johnson LJ, Card SD (2020) Symbiotic bacteria of plant-associated fungi: friends or foes? Curr Opin Plant Biol 56:1–8
2. Hoffman M, Arnold AE (2010) Diverse bacteria inhabit living hyphae of phylogenetically diverse fungal endophytes. Appl Environ Microbiol 76:4063–4075
3. Pawlowska TE (2024) Symbioses between fungi and bacteria: from mechanisms to impacts on biodiversity. Curr Opin Microbiol 80:1–16
4. Julia M, Kelliher JM, Robinson AJ, Longley R, Johnson LYD, Hanson BT, Morales DP, Cailleau G, Junier P, Bonito G, Chain PSG (2023) The endohyphal microbiome: current progress and challenges for scaling down integrative multi-omic microbiome research. Microbiome 11:192
5. Van der Linde K, Lim BT, Rondeel JM, Antonissen LP, de Jong GM (1999) Improved bacteriological surveillance of haemodialysis fluids: a comparison between tryptic soy agar and reasoner's 2A media. Nephrol Dial Transplant 14(10):2433–2437
6. Reasoner DJ, Geldreich EE (1985) A new medium for the enumeration and subculture of bacteria from potable water. Appl Environ Microbiol 49(1):1–7
7. Sandle T (2004) An approach for the reporting of microbiological results from water systems. PDA J Pharm Sci Technol 58(4):231–237
8. Pioli PD (2019) Protocol: Luria-Bertani (LB) media/broth recipe https://med.wmich.edu/sites/default/files/Pioli_Lab_LB_Media_Recipe.pdf. Last accessed on 01.08.2024
9. King EO, Ward MK, Raney DE (1954) Two simple media for the demonstration of pyocyanin and fluorescin. J Lab Clin Med 44:301–307
10. Schaad NW (1980) Laboratory guide for the identification of plant pathogenic bacteria. The American Phytopathological Society, St. Paul, p 72
11. Meenakshi S, Hiremath J, Meenakshi MH, Shivaveerakumar S (2024) Actinomycetes: isolation, cultivation and its active biomolecules. J Pure Appl Microbiol 18(1):118–143
12. Kotb E, Alabdalall AH, Alghamdi AI, Ababutain IM, Aldakeel SA, Al-Zuwaid SK, Algarudi BM, Algarudi SM, Ahmed AA, Albarrag AM (2023) Screening for chitin degrading bacteria in the environment of Saudi Arabia and characterization of the most potent chitinase from *Streptomyces variabilis* Am1. Sci Rep 13:11723
13. Arendt KR, Hockett KL, Araldi-Brondolo SJ, Baltrus DA, Arnold AE (2016) Isolation of endohyphal bacteria from foliar Ascomycota and *in vitro* establishment of their symbiotic associations. Appl Environ Microbiol 82: 2943–2949
14. Aryal S (2022) Tryptic Soy Agar- composition, principle, preparation, results, uses. https://microbenotes.com/tryptic-soy-agar/. Accessed on 10 Aug 2024
15. Bonnet M, Lagier JC, Raoult D, Khelaifia S (2019) Bacterial culture through selective and non-selective conditions: the evolution of culture media in clinical microbiology. New Microbes New Infect 34:10062
16. Gallery RE, Dalling JW, Arnold AE (2007) Diversity, host affinity, and distribution of seed-infecting fungi: a case study with Cecropia. Ecology 88:582–588
17. Lear G, Dickie I, Banks J, Boyer S, Buckley H, Buckley T, Cruickshank R, Dopheide A, Handley K, Hermans S, Kamke J, Lee C, MacDiarmid R, Morales S, Orlovich D, Smissen R, Wood J, Holdaway R (2017) Methods for the extraction, storage, amplification and sequencing of DNA from environmental samples. N Z J Ecol 42:10–50
18. Okrasinska A, Bokus A, Duk K, Gęsiorska A, Sokołowska B, Miłobędzka A, Wrzosek M, Pawłowska J (2021) New endohyphal relationships between Mucoromycota and Burkholderiaceae representatives. Appl Environ Microbiol 87:e02707–e02720
19. Arnold AE, Lutzoni F (2007) Diversity and host range of foliar fungal endophytes: are tropical leaves biodiversity hotspots? Ecology 88:541–549
20. Thomas P, Reddy KM (2013) Microscopic elucidation of abundant endophytic bacteria colonizing the cell wall–plasma membrane perispace in the shoot-tip tissue of banana. AoB Plants 5:plt011
21. Gul F, Hussain A, Jan G, Hamayun M (2017) Genomic DNA extraction for molecular identification of endophytic fungi: an easy and efficient protocol. Biosci Biotech Res Asia 14: 667–672

22. DAPI staining – https://www.biorender.com/template/dapi-staining-protocol-cell-culture. Last accessed on 01 Aug 2024

23. Anonymous (2004) LIVE/DEAD BacLight Bacterial viability kit product information MP 07007.https://assets.thermofisher.com/TFS-Assets/LSG/manuals/mp07007.pdf. Accessed on 15 May 2025

24. Sarkar B, Sultana A, Binti N, Chowdhury F, Afrin S, Fahim M, Rahman T, Rahman A (2024) Nature's pharmacy under siege: investigating antibiotic resistance pattern in endophytic bacteria of medicinal plants. Adv Microbiol 14:183–208

Chapter 2

Isolation and Maintenance of Algal Endosymbionts from Ciliates

Komal Chandarana and Natarajan Amaresan

Abstract

Among protists, ciliates (Ciliophora, Alveolata) are the best candidates for investigation of their endosymbiotic community due to their culturability under laboratory conditions and their presence in a wide variety of environments including ultraoligotrophic niches. However, a lesser number of isolation techniques are reported to date to separate and cultivate the endosymbiont from various ciliate species under lab conditions. This present chapter specifically deals with the isolation of algal endosymbionts from freshwater or marine ciliates. The technique involves micromanipulation of the ciliate, rupturing the cell to release endosymbiotic algae followed by enrichment of endosymbiont on complex media.

Key words Algal endosymbiont, Ciliate, Isolation, Media, Micromanipulation

1 Introduction

Protists, especially ciliate harboring endosymbionts, have been researched since the 1890s [1]. Attempts have been made in the past to isolate, identify, and culture these algal symbionts of ciliate by employing different methods and modifications [2–4]. However, the described methods do not illustrate the isolation of algae from individual protistan cells which was finally proposed by two researchers Achilles-Day and Day [5]. They employed the use of the capillary micromanipulation method under an inverted microscope to draw up a single ciliate cell by capillary action and transfer it to a sterile algal supporting growth medium to culture the algal endosymbiont. This method is described here with some modifications in detail to facilitate the isolation of green endosymbiont of ciliate cell. However, the method can also be employed for other protistan cells like Amoebazoa.

The choice of media will be dependent upon the origin of protists, i.e., freshwater organisms being manipulated in a freshwater-based medium and marine taxa in a marine-based

N. Amaresan and Sapna Chandwani (eds.), *Endosymbionts*, Springer Protocols Handbooks,
https://doi.org/10.1007/978-1-0716-4771-4_2,

medium. Complex media containing natural resources, vitamins, and antibiotics will be more likely to enhance the growth and purification of endosymbiotic algae than defined mineral-based media. Here, we have described two different media to enrich the released algal endosymbionts from protistan cells [6, 7]. However, more different media can be employed to cultivate the algae along with antibiotic amendments.

2 Materials

2.1 Glasswares and Equipment

1. Microcapillary.
2. Microscopic slides.
3. Pasture pipettes.
4. Flat-bottom 12-well microplate.
5. Petri plates.
6. Parafilm.
7. Inverted microscope.

2.2 Media

1. **Composition of Bold's basal medium (BBM) [6]**

Ingredients	g L^{-1}
$NaNO_3$	25.0
$MgSO_4 \cdot 7H2O$	7.50
NaCl	2.50
K_2HPO_4	7.50
KH_2PO_4	17.50
$CaCl_2 \cdot 2H_2O$	2.50
Trace element solution:	1.0 mL
$ZnSO_4 \cdot 7H_2O$	4.42
$MnCl_2 \cdot 4H_2O$	1.44
MoO_3	0.71
$CuSO_4 \cdot 5H_2O$	1.57
$Co(NO_3)_2 \cdot 6H_2O$	0.49
H_3BO_3	11.40
EDTA–KOH solution:	1.0 mL
EDTA–Na_2	50.0
KOH	31.0
$FeSO4 \cdot 7H_2O$ with 1.0 mL concentrated H_2SO_4	4.98

Adjust medium to final pH of 5.5 to 6.5 and autoclave at 121 °C for 20 min

2. **Guillard's f/2 medium composition** [7]

Ingredients	g L^{-1}
$NaNO_3$	0.075
$Na_2HO_4 \cdot 2H2O$	7.50
EDTA–Na_2	4.16
Trace element solution:	1.0 mL
$FeCl_3 \cdot 6H_2O$	3.15
$CoCl_2 \cdot 6H_2O$	0.01
$CuSO_4 \cdot 5H_2O$	1.57
$ZnSO_4 \cdot 7H_2O$	0.022
$MnCl_2 \cdot 4H_2O$	0.180
$Na2MoO4 \cdot 2H_2O$	0.71
Vitamin mix solution:	1.0 mL
Cyanocobalamin (vitamin B12)	0.0005
Thiamine HCl (vitamin B1)	0.10
Biotin	0.0005

3 Methods

3.1 Pre-preparations

1. Prepare all the desired media and sterilize it in advance.
2. Pre-fill the microwell plates with 3.0 mL of pre-sterilized different media.
3. Microcapillary should be heated by a Bunsen burner to make it blunt—creating a "glass-drop" on its one end.
4. Make one end of some microcapillaries as thin as human hair by heating and stretching them over a Bunsen burner, to use it for picking up a single protistan cell.
5. Take a microscopic slide and put four to five drops of pre-sterilized fresh/marine water on it with the help of a Pasteur pipette.

3.2 Preparation of Single Ciliate Cell

1. Take a drop of collected environmental sample onto a microscopic slide and observe under an inverted microscope by performing the wet mount technique.

2. If a moving ciliate is encountered, pick up the single individual cell using a microcapillary while observing under the microscope (*see* **Note 1**).
3. Wash the single ciliate cell several times on a microscopic slide which will be previously seeded with fresh/marine water drops.
4. Transfer the cells into fresh BBM broth and incubate it at an appropriate temperature for 24 h for starvation and digestion of food by ciliates.

3.3 Inoculation, Incubation, and Observation

1. After incubation, again, wash the cell and transfer it onto fresh Guillard's f/2 medium.
2. Before the transfer of ciliates onto agar plates, 50 μL of an antibiotic mix (mixture of 1% penicillin G, 0.25% chloramphenicol, and 0.25% streptomycin) should be added to prevent bacterial growth (*see* **Note 2**).
3. Destroy the ciliate by pressing gently on the agar surface with the help of a microcapillary (*see* **Note 3**).
4. Transfer the green cells released, using a micropipette to a 12-well flat bottom microtiter plate, aseptically. Inoculate the algae in three wells pre-seeded with 3 mL of different media.
5. Seel the plates with parafilm and incubate them at 20 °C in 12: 12-h dark:light conditions for approximately 30 days (*see* **Note 4**).
6. If the algal growth is observed in the plates, streak it onto agar plates, employing the same medium solidified by the addition of 15 g L^{-1} agar.
7. Incubate the Petri plates for 7–28 days under standard conditions and observe the development of microcolonies.
8. Transfer the discrete colonies to Petri plates containing the appropriate medium with the help of pre-sterilize tooth-sticks and incubate as above (*see* **Note 5**).
9. The stock cultures can be maintained as liquid cultures or on agar slants under standard environmental conditions. However, it should be routinely transferred every 2–3 months.

4 Notes

1. The person should be very efficient and experienced in picking up the single moving cell.
2. One can confirm the single cell under the microscope at this stage; if not, then one can again pick up the single cell from this plate and transfer it onto a fresh plate.

3. The pressing will make a small dent on the agar while destroying the ciliate cell. One must ensure that the cell is ruptured releasing algal endosymbiont.
4. It may be necessary to adjust incubation regimes for samples derived from different environmental conditions. It is advisable to observe the incubated plates at regular 5–7-day intervals until the appearance of green algal growth.
5. It is necessary to obtain pure cultures. If necessary, repeat **steps 7–9** to ensure clonal cultures.

References

1. Beijerinck MW (1890) Cultivation experiments with zoochlorella, lichenengonidia and other lower algae. Bot Ztg 48:781–788
2. Reisser W (1981) The endosymbiotic unit of *Stentor polymorphus* and *Chlorella* sp. morphological and physiological studies. Protoplasma 105:273–284
3. Nishihara N, Horiike S, Takahashi T, Kosaka T, Shigenaka Y, Hosoya H (1998) Cloning and characterization of endosymbiotic algae isolated from *Paramecium bursaria*. Protoplasma 203: 91–99
4. Kamako SI, Hoshina R, Ueno S, Imamura N (2005) Establishment of axenic endosymbiotic strains of Japanese *Paramecium bursaria* and the utilization of carbohydrate and nitrogen compounds by the isolated algae. Eur J Protistol 41:193–202
5. Achilles-Day UE, Day JG (2013) Isolation of clonal cultures of endosymbiotic green algae from their ciliate hosts. J Microbiol Methods 92:355–357
6. Khan ANMAI, Habib MAB, Hossain MS, Miah MI (2018) Culture of Chlorella vulgaris in press mud media as sugar mill waste. Culture 3:41–45
7. García L, Álvarez X, Bravo K, Peralta J, Barriga A (2012) Obtaining microalgae biomass in a small-scale reactor. RE & PQJ 10:1492–1494

Chapter 3

Isolation and Maintenance of Methanogenic Endosymbionts

Sarika Kumari and Natarajan Amaresan

Abstract

This chapter demonstrates a step-by-step and detailed protocol for isolating and culturing methanogenic endosymbionts which are crucial organisms in aid of digestion and methane synthesis in different niches. The first step of the approach is sample collection from anaerobic habitats through the enumeration technique and then the selection of specialized culture media that favors the methanogenic genera while adversely affecting other microbes. In this book chapter, we have described general methods of maintaining an anaerobic environment for the methanogens such as the use of gas-tight containers and a specific gaseous environment to enhance their growth. Key characteristics of these organisms' isolation and long-term cultivation include the temperature, pH, nutrient content, and the need for some trace elements. Thus, this chapter intends to provide a reference to the investigators concerned with the ecological and biotechnological aspects of methanogenic endosymbionts. The isolation and maintenance of these microorganisms could be utilized effectively in the sustainable disposal of wastes, generation of renewable energies, and replenishment and conservation of the environment.

Key words Methanogenic, Endosymbionts, Isolation, Maintenance, Anaerobic digestion

1 Introduction

Methanogenic endosymbionts can be defined as a separate group of anaerobic microbiotas that have critical importance to methane production. These organisms are anaerobic and mostly parasitic or symbiotic with some protozoa, insects, and higher animals in the digestion of highly composed organic substrates [1]. They cannot just be methane-producing buggers; they are the most critical inhabitants in many environments that break down organic compounds and recycle nutrients/energy [2, 3].

In the course of methanogenesis, these endosymbionts assist in the fermentation of plants and animals producing methane and carbon dioxide. It also assists in the reduction of the emission of greenhouse gases; at the same time, biogas can be captured from the process for use as an energy source. Methanogenic endosymbionts also have a role in the biogeochemical cycles—carbon and

N. Amaresan and Sapna Chandwani (eds.), *Endosymbionts*, Springer Protocols Handbooks,
https://doi.org/10.1007/978-1-0716-4771-4_3,

nitrogen—as well as elemental flow in various ecosystems such as the wetland, ruminant's gut, and municipal solid waste sludge digesters [4, 5]. Their capability for substrate-level phosphorylation enables them to harness nutrients that would be otherwise inaccessible to most other organisms; this points to their vital role in the overall global metabolism [6]. Therefore, studies on methanogenic endosymbionts are essential for enhancing knowledge of biotechnological utilizations and addressing important questions concerning waste disposal and climate change.

2 Materials and Media Required for the Isolation of Methanogenic Endosymbionts

2.1 Isolation Medium for Methanogenic Endosymbionts

The medium contains 32 g/L artificial sea salts, 1.2 g/L $Na_2S.9H_2O$, and 0.5 g/L Na_2CO_3. Optional components include 20 g/L neutralized liver digest, 0.4 g/L ground dried grass or 2 g/L lactate, and 2 g/L yeast extract. It also includes 2 mL of trace minerals solution, 20 mL of vitamin solution, and 10% heat-inactivated sterile calf serum.

Preparation

1. Dissolve the specified amounts of artificial sea salts, $Na_2S{\cdot}9H_2O$, Na_2CO_3, and any optional components in a sterile beaker, then add distilled water to reach a total volume of 1 liter.
2. Measure the pH of the solution using a pH meter and adjust it to 7.0–7.4 with dilute HCl or NaOH as necessary [7].
3. Autoclave the mixture at 121 °C for 15–20 min to ensure sterility, then allow it to cool under sterile conditions (*see* **Note 1**).
4. Once cooled, add heat-inactivated sterile calf serum to achieve a final concentration of 10%, along with trace minerals and vitamin solutions [8].
5. Store the prepared medium in sealed containers under anaerobic conditions, such as nitrogen gas, until use (*see* **Note 2**).

2.2 Phosphate Buffer Preparation

To prepare phosphate buffer, dissolve 1.42 g of sodium phosphate dibasic (Na_2HPO_4) and 0.24 g of sodium phosphate monobasic (NaH_2PO_4) in distilled water. Mix thoroughly and make up the volume to 1 liter with distilled water. This buffer helps maintain a stable pH and is commonly used in various biological and chemical experiments.

Preparation Steps

1. Weigh out the appropriate amounts of dibasic and monobasic sodium phosphate, then add them to a beaker containing distilled water, and stir until fully dissolved.

2. Check the solution's pH and adjust it to 7.0 using small amounts of HCl or NaOH as needed.
3. Autoclave the buffer solution at 121 °C for 15 min to sterilize it.
4. Store the buffer solution in sterile containers at room temperature or refrigerate as required.

2.3 Preparation of Trace Mineral Solution for Methanogenic Endosymbionts [9]

The trace mineral solution for methanogenic endosymbionts is prepared by dissolving specific micronutrients per liter of distilled water. It includes 0.1 g of $FeCl_3.6H_2O$, 0.01 g each of MnCl2.4H2O and $ZnSO_4.7H_2O$, and 0.001 g each of $CoCl_2.6H_2O$, $NiCl_2.6H_2O$, $CuSO_4.5H_2O$, and $Na_2MoO_4.2H_2O$. This solution provides essential trace elements required for optimal growth and metabolism of methanogenic microbes.

Preparation

1. Weigh out the specified amounts of each trace mineral, then add them to a sterile beaker containing distilled water, and stir until fully dissolved.
2. Ensure the total volume reaches 1 liter by adding more distilled water if necessary (*see* **Note 3**).

Transfer the solution to an appropriate container, such as a glass bottle, and autoclave it at 121 °C for 15 min to sterilize. Allow the solution to cool down and store it in sterile conditions, preferably in the dark to prevent degradation of the components.

3 Methods

3.1 Sample Collection

- Collect samples from environments such as anaerobic sediments, marshes, or the gastrointestinal tracts of animals.
- Ensure samples are kept anaerobic during collection and transport.
- Use a pipette to collect cells (e.g., *Metopus contortus* or *Pelomyxa palustris*) [10].
- Carefully transfer the cells to avoid damaging them, ensuring minimal disturbance to their environment.

3.2 Washing

- After collection, wash the cells in a 10 mM phosphate buffer (pH 7.0) to remove any contaminants. This step is crucial to ensure that the isolated endosymbionts are not influenced by external microorganisms.

3.3 Transfer to Anaerobic Environment

- Move the washed cells into an anaerobic glove box to maintain a suitable environment for anaerobic organisms [11].

3.4 Further Isolation

- From the anaerobic environment, cells can be transferred to the isolation medium where they will be ruptured out to release the endosymbionts.

3.5 Filtration (If Applicable)

- If the initial culture contains flocculated materials, filtering through a fine mesh (e.g., 100-μm filter gauze) can be performed to collect clean cells.

3.6 Protocol for the Isolation of Methanogenic Endosymbionts

1. Collect the cells using a pipette.
2. Wash the collected cells in 10 mM phosphate buffer (pH 7.0) to remove contaminants.
3. Move the washed cells into an anaerobic glove box to maintain an oxygen-free atmosphere.
4. Transfer the washed cells to 100 μL of the prepared isolation medium and use a syringe to rupture the cell, releasing its contents (*see* **Note 4**).
5. Fill a glass prefilled tube with the resulting suspension, seal the ends by melting them in a flame, and attach the tube to a microscope slide for incubation at 37 °C [12].
6. Monitor the growth by determining the number of thick endosymbionts (thick rods) in a defined region of the suspension at several time points during incubation, and note down their characteristic shape (*see* **Notes 5** and **6**) [13, 14].

3.7 Maintenance of Methanogenic Endosymbionts

Maintaining methanogenic endosymbionts requires careful attention to their anaerobic environment, nutrient supply, and overall culture conditions. Here are the key steps:

3.7.1 Culture Conditions

- Anaerobic environment: Always maintain cultures in an anaerobic chamber, or use anaerobic jars flushed with an inert gas (e.g., nitrogen or a mixture of H_2/CO_2) [15].
- Temperature control: Incubate cultures at optimal temperatures (typically around 20–37 °C, depending on the specific methanogens) [16].

3.7.2 Medium Replacement

- Subculturing: Transfer a portion of the culture to fresh medium at 1–2-week intervals to prevent nutrient depletion and accumulation of toxic byproducts.
- Volume replacement: During maintenance, remove approximately 50% of the culture volume and replace it with a fresh anaerobic medium.

3.7.3 Nutrient Supply

- Feeding: Regularly add organic substrates such as acetate, formate, or other relevant nutrients, ensuring that the concentration is appropriate for the growth of methanogenic endosymbionts [17].
- Yeast extract and trace elements: Supplement the medium with yeast extract and vitamins to support the growth and metabolism of endosymbionts [18].

3.7.4 Monitoring

- Cell viability: Regularly monitor the culture under a microscope to assess the health and density of the endosymbionts.
- Methane production: Methane production is measured as an indirect indicator of metabolic activity and the health of the culture [19].

3.7.5 Contamination Checks

- Regular inspections: During routine monitoring, check for any signs of contamination (e.g., unwanted microbial growth). If contamination is detected, take appropriate action to isolate healthy cultures [20].
- Use of antibiotics: If necessary, antibiotics (e.g., penicillin) could be included in the medium to suppress unwanted microbial growth [16].

3.7.6 Storage

- Long-term storage: For long-term preservation, consider cryopreservation techniques. Freeze cultures in glycerol or DMSO at −80 °C or in liquid nitrogen [21].
- Periodic revival: Revive frozen stocks periodically to ensure viability and maintain genetic integrity over time.

4 Notes

1. Sterility: Maintain aseptic techniques throughout the preparation to prevent contamination.
2. Anaerobic conditions: For any media that will be used for culturing methanogens, ensure they are stored and handled in anaerobic conditions to support the growth of anaerobic organisms.
3. Final concentration: Adjust the amount used in your main media according to the specific requirements of your experiments. Typically, the final concentration in the growth medium is around 1–2 mL of this solution per liter of culture medium.
4. Cell handling: Handle cells gently to avoid damaging them during transfer and washing.

5. Monitoring conditions: Regularly check incubation conditions (temperature, anaerobic environment) to ensure optimal growth of the methanogenic endosymbionts.
6. Microscopy: Employ appropriate microscopic techniques (e.g., Leitz epifluorescence microscopy) to visualize and identify the endosymbionts effectively.

References

1. Schwarz MJ, Frenzel P (2005) Methanogenic symbionts of anaerobic ciliates and their contribution to methanogenesis in an anoxic rice field soil. FEMS Microbiol Ecol 52(1):93–99
2. Offre P, Spang A, Schleper C (2013) Archaea in biogeochemical cycles. Ann Rev Microbiol 67(1):437–457
3. Krzmarzick MJ, Taylor DK, Fu X, McCutchan AL (2018) Diversity and niche of archaea in bioremediation. Archaea 2018(1):3194108
4. Sugimoto A, Inoue T, Tayasu I, Miller L, Takeichi S, Abe T (1998) Methane and hydrogen production in a termite-symbiont system. Ecol Res 13:241–257
5. Torres-Alvarado R, Ramírez-Vives F, Fernández FJ, Barriga-Sosa I (2005) Methanogenesis and methane oxidation in wetlands: implications in the global carbon cycle. Hidrobiológica 15(3):327–349
6. Lancaster JR Jr (1986) A unified scheme for carbon and electron flow coupled to ATP synthesis by substrate-level phosphorylation in the methanogenic bacteria. FEBS Lett 199(1): 12–18
7. Edwards T, McBride BC (1975) New method for the isolation and identification of methanogenic bacteria. Appl Microbiol 29(4):540–545
8. Wolin EA, Wolin M, Wolfe RS (1963) Formation of methane by bacterial extracts. J Biol Chem 238(8):2882–2886
9. van Bruggen JJ, Zwart KB, van Assema RM, Stumm CK, Vogels GG (1984) Methanobacterium formicicum, an endosymbiont of the anaerobic ciliate Metopus striatus McMurrich. Arch Microbiol 139:1–7
10. van Bruggen JJ, Stumm CK, Vogels GD (1983) Symbiosis of methanogenic bacteria and sapropelic protozoa. Arch Microbiol 136: 89–95
11. Shinzato N, Kamagata Y (2010) The methanogenic and eubacterial endosymbionts of trimyema. In: Hackstein J (ed) (Endo)-symbiotic methanogenic archaea, Microbiology monographs, vol 19. Springer, Berlin/ Heidelberg. https://doi.org/10.1007/978-3-642-13615-3_4
12. Perfil'ev BV, Gabe DR, Shewan JM (1969) Capillary methods of investigating microorganisms. Oliver & Boyd, Edinburgh
13. Méndez-Sánchez D, Schrecengost A, Rotterová J, Koštířová K, Beinart RA, Čepička I (2024) Methanogenic symbionts of anaerobic ciliates are host and habitat specific. ISME J 18(1):wrae164
14. Inoue JI, Noda S, Hongoh Y, Ui S, Ohkuma M (2008) Identification of endosymbiotic methanogen and ectosymbiotic spirochetes of gut protists of the termite Coptotermes formosanus. Microbes Environ 23(1):94–97
15. Hedderich R, Whitman WB (2006) Physiology and biochemistry of the methane-producing archaea. In: The prokaryotes, vol 2, pp 1050–1079
16. Joblin KN (2005) Methanogenic archaea. In: Methods in gut microbial ecology for ruminants. Springer, Dordrecht, pp 47–53
17. Jarrell KF, Kalmokoff ML (1988) Nutritional requirements of the methanogenic archaebacteria. Can J Microbiol 34(5):557–576
18. Wintsche B, Glaser K, Sträuber H, Centler F, Liebetrau J, Harms H, Kleinsteuber S (2016) Trace elements induce predominance among methanogenic activity in anaerobic digestion. Front Microbiol 7:2034
19. Holmes DE, Giloteaux L, Orellana R, Williams KH, Robbins MJ, Lovley DR (2014) Methane production from protozoan endosymbionts following stimulation of microbial metabolism within subsurface sediments. Front Microbiol 5:366
20. Edwards AN, Suárez JM, McBride SM (2013) Culturing and maintaining Clostridium difficile in an anaerobic environment. J Vis Exp 79:e50787
21. Smirnova DV, Zalomova LV, Zagainova AV, Makarov VV, Mezhevikina LM, Fesenko EE Jr, Yudin SM (2019) Cryopreservation of the human gut microbiota: current state and perspectives. Int J Med Microbiol 309(5): 259–269

Chapter 4

Isolation and Maintenance of Cyanobacterial Endosymbionts

Bhavana V. Mohite, Sunil H. Koli, and Satish V. Patil

Abstract

Cyanobacteria are a wide and diversified phylum of bacteria capable of photosynthesis. Cyanobacteria are the pioneer colonizers of a particular environment, and establishment in the environment is preceded by microbial consortia formation. Some heterotrophic bacteria are in symbiotic association with cyanobacteria for the carbon and nitrogen exchange and recycling of the key nutrients. The heterotrophic bacteria are coexisting partners with cyanobacteria in the cyanosphere. When cyanobacteria are in consortium with heterotrophic bacteria, the system becomes more stable with a longer lifespan. The present protocol focused on the isolation of endosymbiont heterotrophic bacteria of cyanobacteria in brief.

Key words Cyanosphere, Heterotrophic bacteria, Consortium, Interspecies interaction

1 Introduction

Cyanobacteria constitute pioneer colonizers of specific environments, and their settlement in new biotopes precedes the establishment of intricate microbial consortia [1, 2]. Indeed, these autotrophic microorganisms rarely live in isolated populations but thrive in complex communities where intricate interactions take place [3]. Cyanobacteria are frequently associated with heterotrophic bacteria in nature [4], and geological evidence of interspecies interactions between cyanobacteria and heterotrophs dates to 440 million years ago [5]. The cohesiveness of such interactions is illustrated by the paucity of axenic cyanobacterial strains, despite persistent attempts to isolate them from their consortiums [6].

Cultures of isolated cyanobacterial strains typically contain coexisting heterotrophic bacterial partners that are considered their phycosphere, the so-called cyanosphere [7]. Although axenic cultures of cyanobacteria can be achieved, cyanobacterial cultures comprising associated coexisting bacterial populations are more

N. Amaresan and Sapna Chandwani (eds.), *Endosymbionts*, Springer Protocols Handbooks,
https://doi.org/10.1007/978-1-0716-4771-4_4,

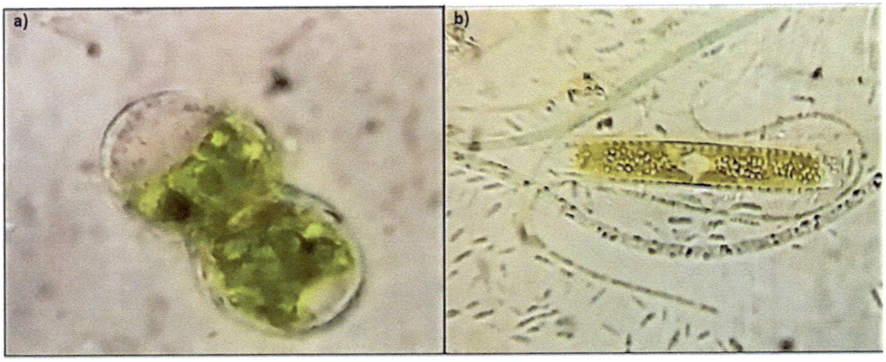

Fig. 1 Representative images of endosymbionts of (**a**) cyanobacteria and (**b**) diatoms (at Dr. Satish V Patil, Laboratory, KBC North Maharashtra University, Jalgaon)

stable and have more robust and longer lifespans than axenic ones [8].

Numerous heterotrophic bacteria are associated with cyanobacteria and have a substantial influence on cyanobacterial growth (Fig. 1) [9–12].

A unicyanobacterial consortium (UCC) is a phototrophic community in which one such cyanobacterium, serving as the sole or primary autotroph, supplies photosynthetically derived carbon and oxygen to one or more heterotrophs and plays an essential role in the aquatic food web [13]. The heterotrophs, in turn, can promote cyanobacterial growth by providing key metabolites and scavenging wastes [14, 15]. In this way, a consortium's expanded and compartmentalized metabolic capacity synergistically improves resource utilization efficiency over that of its individual members [16].

Cyanobacteria and their endosymbiont heterotrophic bacteria consortium may have physical interaction for carbon and nitrogen exchange such as between *Anabaena* sp. SSM-00 and *Rhizobium* sp. WH2K [17]. Many cyanobacteria grow more efficiently in co-cultivation with heterotrophic bacteria than in axenic cultures, especially when heterotrophic remineralization of organic carbon alleviates cyanobacterial carbon limitation [18].

Cyanobacterial–heterotrophic bacterial associations can be influenced by abiotic factors including nutrients, light, temperature conditions, pH, and salinity as well as biotic factors like grazing [11]. Few heterotrophic bacterial taxa appeared remarkably shared among different cyanospheres (belonging to proteobacterial orders: Burkholderiales, Rhizobiales, Sphingomonadales, and Caulobacteriales) and not to be the most dominant of the mud biofilms, suggesting host-specific association with each cyanobacterial taxon in culture [7]. In a recent study, it was demonstrated that *M. vaginatus* has a unique bacterial community associated with its bundles, a microscopic "cyanosphere" (by analogy to the rhizosphere). Not only does this cyanosphere community contain heterotrophic nitrogen-fixing bacteria, but its nitrogen fixation potential is typically 100-fold higher than that of the surrounding biocrust soil [19]. Such spatial concentration of nitrogen-fixing heterotrophs suggests a potential resource trading relationship where fixed carbon is traded for fixed nitrogen.

2 Materials

1.1 Marine-polluted cyanobacterial mat.

1.2 Defined artificial seawater medium [20] or BG11 Medium [21].

1.2.1 $MgCl_2.6H_2O$ (5.6 g/L), $MgSO_4.7H_2O$ (6.8 g/L), $CaCl_2.2H_2O$ (1.47 g/L), KCl (0.66 g/L), and KBr (0.09 g/L). After autoclaving, add 0.15 g/L and 0.2 g/L solutions of KH_2PO_4 and NH_4Cl.

1.2.2 Add 1 mL autoclaved solutions of trace elements [(-$Na_2EDTA.2H_2O$ (4.36 g/L), $FeCl_3{\cdot}6H_2O$ (3.15 g/L), $MnCl_2{\cdot}4H_2O$ (1 mL of primary stock 178.10 g/L), $NiSO_4{\cdot}6H_2O$ (1 mL of primary stock 2.63 g/L), $ZnSO_4{\cdot}7H_2O$ (1 mL of primary stock 23.00 g/L), $CoCl_2{\cdot}6H_2O$ (1 mL of primary stock 11.90 g/L), $CuSO_4{\cdot}5H_2O$ (1 mL of primary stock 2.50 g/L), $Na_2MoO_4{\cdot}2H_2O$ (19.90 g/L), $H2SeO_3$ (1.29 g/L), Na_3VO_4 (1.84 g/L), K_2CrO (1.94 g/L). Bring the final volume to 1 L with distilled water and filter sterilize the final trace element solution through a 0.2-μ PES membrane].

1.2.3 Add 1 mL of vitamin solution [200 mg thiamine HCl (vit. B1), 10 mL primary stock solution of biotin (vit. H) (0.1 g/L), 1 mL primary stock solution of cyanocobalamin (vit. B12) (1.0 g/L). Bring the final volume to 1 L with distilled water and filter sterilize through a 0.2-μ PES membrane].

1.3 Add bacteriological agar [1% (w/v)].

1.4 Adjust the salinity of the medium to 2% (*see* **Note 1**).

1.5 10 mM TES buffer (pH 8.0).

1.6 Screw cap bottles (20 mL capacity).

1.7 Sterile distilled water.

3 Methods

3.1 Isolation of Aerobic Endosymbiont Heterotrophic Bacteria

1. Prepare the defined artificial seawater medium in a 20-mL screw-cap culture tube containing 10 mL culture medium under aerobic conditions [20]. Inoculate the cyanobacterial mat in it for enrichment.
2. Prepare a serial dilution of the cyanobacterial mat from the enriched culture. Homogenize the cyanobacterial mat and mix it with sterile 9 mL of 10 mM TES (pH 8.0) buffer to prepare the 10^{-1} dilution. Prepare the further dilutions up to 10^{-6} by transferring 1 mL from the previous dilution tube.
3. Inoculate 0.1 mL of inoculum from each of the dilution tubes onto solid defined artificial seawater medium, perform the spread plate method and incubate the plates at room temperature (Fig. 2).
4. Isolate pure culture by picking up the single colony and streak the bacterial colony at least five times to obtain pure heterotroph cultures. Confirm the presence of cyanobacterial heterotrophic bacteria primarily on nutrient agar and some specific media such as BG-11, nitrogen-free media, and MacConkey agar and also perform the gram staining technique [16].
5. Further assess the purity of the isolates by light microscopy or electron microscopy and 16S rRNA gene sequencing [16].
6. Maintain the heterotrophic bacteria thereafter by subculturing at 4-week intervals on fresh defined artificial seawater medium.
7. Further assessment of the specific heterotrophic bacteria could be carried out by using selective media (*see* **Note 2**).

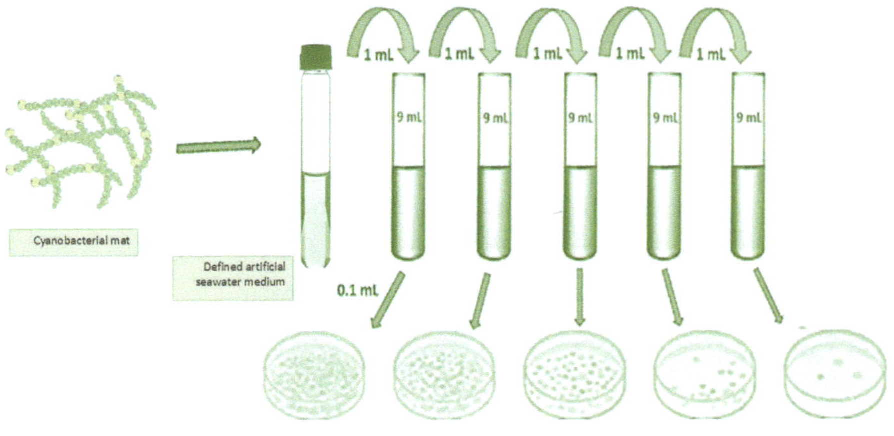

Fig. 2 Isolation of heterotrophic bacteria from cyanobacterial mat

4 Notes

1. The maintenance of the salinity of the medium should be similar to the salinity of the field site where the strains originated.
2. The selective heterotrophic bacteria could be isolated after growth on defined artificial marine medium /BG11 by enrichment of cyanobacterial exudates as the sole carbon source like fatty acids, alcohols, amino acids, and sugars containing medium. For example, hydrocarbon-degrading bacteria could be enriched in crude oil and further using selective media such as acetate/alkane-containing medium [20]. Nitrogen-fixing isolates could be confirmed by checking their growth on nitrogen free medium and Rhizobium like isolate could be confirmed by growing on yeast extract mannitol agar (YEMA) medium [22] and by nodulations test [23].

Acknowledgments

Author BVM is grateful to the principal of Bajaj College of Science, Wardha, for the encouragement and support.

References

1. Whitton BA (ed) (2012) Ecology of cyanobacteria II: their diversity in space and time. Springer, Dordrecht
2. Mazard S, Penesyan A, Ostrowski M, Paulsen I, Egan S (2016) Tiny microbes with a big impact: the role of cyanobacteria and their metabolites in shaping our future. Mar Drugs 14:97
3. Seymour JR, Amin SA, Raina JB, Stocker R (2017) Zooming in on the phycosphere: the ecological interface for phytoplankton– bacteria relationships. Nat Microbiol 2:1–12
4. Paerl HW (1977) Role of heterotrophic bacteria in promoting N2 fixation by anabaena in aquatic habitats. Microb Ecol 4:215–231
5. Tomescu AMF, Honegger R, Rothwell GW (2008) Earliest fossil record of bacterial–cyanobacterial mat consortia: the early Silurian Passage Creek biota (440 Ma, Virginia, USA). Geobiology 6:120–124
6. Ferris MJ, Hirsch CF (1991) Method for isolation and purification of cyanobacteria. Appl Environ Microbiol 57:1448–1452
7. Halary S, Duperron S, Demay J, Duval C, Hamlaoui S, Piquet B et al (2022) Metagenome-based exploration of bacterial communities associated with cyanobacteria strains isolated from thermal muds. Microorganisms 10:2337
8. Jackrel SL, Yang JW, Schmidt KC, Denef VJ (2021) Host specificity of microbiome assembly and its fitness effects in phytoplankton. ISME J 15:774–788
9. Eiler A, Bertilsson S (2004) Composition of freshwater bacterial communities associated with cyanobacterial blooms in four Swedish lakes. Environ Microbiol 6:1228–1243
10. Kolmonen E, Sivonen K, Rapala J, Haukka K (2004) Diversity of cyanobacteria and heterotrophic bacteria in cyanobacterial blooms in Lake Joutikas, Finland. Aquat Microbiol Ecol 36:201–211
11. Dziallas C, Grossart HP (2011) Temperature and biotic factors influence bacterial communities associated with the cyanobacterium Microcystis sp. Environ Microbiol 13:1632–1641
12. Adam B, Klawonn I, Svedén JB, Bergkvist J, Nahar N, Walve J et al (2016) N2-fixation, ammonium release and N-transfer to the microbial and classical food web within a plankton community. ISME J 10:450–459

13. Woodhouse JN, Kinsela AS, Collins RN, Bowling LC, Honeyman GL, Holliday JK et al (2016) Microbial communities reflect temporal changes in cyanobacterial composition in a shallow ephemeral freshwater lake. ISME J 10: 1337–1351
14. Paerl HW, Pinckney JL (1996) A mini-review of microbial consortia: their roles in aquatic production and biogeochemical cycling. Microb Ecol 31:225–247
15. Paerl HW, Pinckney JL, Steppe TF (2000) Cyanobacterial–bacterial mat consortia: examining the functional unit of microbial survival and growth in extreme environments. Environ Microbiol 2:11–26
16. Cole JK, Hutchison JR, Renslow RS, Kim Y-M, Chrisler WB, Engelmann HE et al (2014) Phototrophic biofilm assembly in microbial-mat-derived unicyanobacterial consortia: model systems for the study of autotroph-heterotroph interactions. Front Microbiol 5: 109
17. Behrens S, Lösekann T, Pett-Ridge J, Weber PK, Ng W-O, Stevenson BS et al (2008) Linking microbial phylogeny to metabolic activity at the single-cell level by using enhanced element labeling-catalyzed reporter deposition fluorescence in situ hybridization (EL-FISH) and NanoSIMS. Appl Environ Microbiol 74: 3143–3150
18. Shen H, Niu Y, Xie P, Tao M, Yang X (2011) Morphological and physiological changes in Microcystis aeruginosa as a result of interactions with heterotrophic bacteria. Freshw Biol 56:1065–1080
19. Couradeau E, Giraldo-Silva A, De Martini F, Garcia-Pichel F (2019) Spatial segregation of the biological soil crust microbiome around its foundational cyanobacterium, microcoleus vaginatus, and the formation of a nitrogen-fixing cyanosphere. Microbiome 7:111–122
20. Abed RMM (2010) Interaction between cyanobacteria and aerobic heterotrophic bacteria in the degradation of hydrocarbons. Int Biodeterior Biodegrad 64:58–64
21. Allen MM, Stanier RY (1968) Growth and Division of Some Unicellular Blue-green Algae. J Gen Microbiol 51:199–202
22. Zheng Q, Hu Y, Kosina SM, Van Goethem MW, Tringe SG, Bowen BP et al (2023) Conservation of beneficial microbes between the rhizosphere and the cyanosphere. New Phytol 240:1246–1258
23. Somasegaran P, Hoben HJ (1994) Handbook for Rhziobia: Methods in legume-Rhizobium technology. Springer-Verlag Publisher. https://doi.org/10.1007/978-1-4613-8375

Chapter 5

Isolation of Fungal Endosymbionts

T. S. Archana, Devendra Kumar, Shivam Singh, and Gagan Kumar

Abstract

Isolating fungal endosymbionts is crucial for understanding their roles in various ecosystems, agriculture, and biotechnology. Fungal endosymbionts residing within plant tissues play essential roles in enhancing plant growth, stress tolerance, and disease resistance. However, their isolation presents several challenges, including host specificity, non-culturable nature, and contamination risks. Traditional methods, such as morphological identification, are often supplemented or replaced with molecular techniques, including DNA sequencing of conserved regions like the ITS and 18S rRNA genes, which provide accurate identification of endosymbionts. Recent advancements have enabled the isolation of diverse endosymbiotic fungi, such as *Epichloë festucae*, *Piriformospora indica*, *Trichoderma harzianum*, *Fusarium oxysporum*, and *Penicillium chrysogenum*, each of which offers unique benefits, ranging from biocontrol to antibiotic production. This book chapter highlights the importance of combining traditional and molecular approaches to overcome the challenges of isolation, ensuring a deeper understanding of fungal endosymbionts and their potential applications across various fields.

Key words Endosymbionts, Isolation, Techniques

1 Introduction

Fungal endosymbionts play a critical role in the dynamics of host organisms and their environments, often engaging in mutualistic relationships where both parties benefit each other. In plants, for example, mycorrhizal fungi enhance nutrient uptake, particularly phosphorus, and protect against soil pathogens while receiving carbohydrates from the plant [1, 2]. In animals, fungal endosymbionts can aid digestion and detoxification and boost immune responses. These relationships are essential not only for individual host organisms but also for broader ecological processes such as soil fertility and ecosystem stability [3]. Isolating fungal endosymbionts is vital for studying these interactions. Through isolation and cultivation, researchers can explore the fungi's physiological traits, genetic makeup, and host interactions under controlled conditions. This understanding is especially valuable in agriculture for

N. Amaresan and Sapna Chandwani (eds.), *Endosymbionts*, Springer Protocols Handbooks,
https://doi.org/10.1007/978-1-0716-4771-4_5,

developing sustainable practices, such as using fungal endosymbionts to improve crop growth, reduce chemical fertilizer dependency, and enhance resistance to pests and diseases [4]. Additionally, in environmental science, these fungi are crucial in bioremediation, helping to break down pollutants and restore contaminated environments, while in medicine, they offer potential sources for new antimicrobial compounds to combat antibiotic resistance [5]. Furthermore, studying fungal endosymbionts deepens our understanding of microbial diversity and evolution, shedding light on how symbiotic relationships have evolved and how they stabilize ecosystems under environmental stress. Thus, isolating fungal endosymbionts is not only practical but also provides insights into fundamental biological processes that sustain life on Earth [6].

Fungal endosymbionts can be isolated from various sources, each revealing unique aspects of their symbiotic relationships. Plant tissues, such as roots, leaves, stems, and seeds, are common sources. Roots, in particular, are rich in symbiotic fungi like mycorrhizal fungi, which enhance nutrient uptake [7]. Leaves and stems may contain endophytic fungi that protect against herbivores and pathogens, while seeds can harbor fungi aiding in germination and early development. Invertebrates, including insects and nematodes, also host fungal endosymbionts that are crucial for their biology, such as aiding digestion or protecting against pathogens. For instance, some insects rely on fungal endosymbionts to break down complex plant materials essential for their survival [8–12].

2 Material

2.1 Surface Sterilization

1. Plant or animal tissues (e.g., leaves, roots, stems)
2. Sterile containers or bags for transport
3. Sterile water for rinsing
4. 70% ethanol for surface disinfection (*see* **Note 1**)
5. 2–5% sodium hypochlorite solution (NaOCl) (*see* **Note 2**) for further disinfection
6. Sterile filter paper or paper towels for drying (*see* **Note 3**)
7. Sterile nutrient agar plates (e.g., potato dextrose agar) for culturing
8. Sterile forceps for handling tissue (*see* **Note 4**)
9. Sterile scissors or scalpel for cutting tissue (*see* **Note 5**)
10. Incubator set to 25–28 °C

2.2 Tissue Fragmentation and Plating

1. Healthy plant tissues (e.g., leaves, roots, stems) (*see* **Note 6**)
2. Sterile scissors or scalpel for cutting tissue
3. Sterile Petri dishes for culturing
4. Sterile culture media (e.g., potato dextrose agar, malt extract agar) (*see* **Note 7**)
5. Sterile forceps for placing tissue fragments
6. Parafilm or other sealing material for Petri dishes (*see* **Note 8**)
7. Incubator set to 25–28 °C
8. Sterile loop or needle for subculturing
9. Sterile filter paper or paper towels for blotting

2.3 Serial Dilution and Plating

1. Plant tissue samples (e.g., leaves, roots) (*see* **Note 9**)
2. Sterile test tubes and pipettes
3. Sterile water or buffer for dilutions
4. Sterile mortar or blender for homogenizing tissue (*see* **Note 10**)
5. Sterile culture media (e.g., potato dextrose agar)
6. Sterile spreader or glass rod for plating
7. Incubator set to 25–28 °C (*see* **Notes 11, 12**)
8. Sterile loop or needle for subculturing

2.4 Single Spore Isolation

1. Fungal cultures with sporulating colonies
2. Sterile Petri dishes with culture media (e.g., potato dextrose agar)
3. Sterile loop or needle for picking spores (*see* **Note 13**)
4. Dissecting microscope for visualizing spores(*see* **Note 14**)
5. Sterile water or buffer for spore suspension (*see* **Note 15**)
6. Parafilm or other sealing material for Petri dishes
7. Incubator set to 25–28 °C

2.5 Symbiotic Culture Techniques

1. Host tissues (e.g., plant roots, insect guts)
2. Sterile Petri dishes and culture media (e.g., water agar)
3. Sterile water or buffer for spore suspension (*see* **Note 15**)
4. Fine needles or dissecting tools for handling fungi
5. Sterile pipettes for transferring
6. Incubator set to 25–28 °C (*see* **Note 16**)
7. Sterile forceps and dissecting microscope for precise handling

2.6 Culture-Independent Methods

1. Healthy plant tissues (e.g., leaves, roots, stems)
2. Sterile containers for tissue transport
3. Sterile water for rinsing
4. Mortar and pestle or mechanical tissue homogenizer for homogenization
5. DNA extraction kit for plant tissues
6. Spectrophotometer or fluorometer for DNA quantification
7. PCR reagents and primers for fungal DNA amplification
8. Gel electrophoresis equipment for PCR product analysis
9. Sequencing library preparation kit and sequencing platform (e.g., Illumina)
10. Bioinformatics tools for data analysis

2.7 Microscopy-Guided Isolation

1. Healthy plant tissues (e.g., leaves, roots, stems) (*see* **Note 17**)
2. Sterile containers for tissue transport
3. Sterile water for rinsing
4. Cryostat or microtome for sectioning tissues.
5. Adhesive-coated slides compatible with LCM
6. Histological stains or fluorescent dyes (optional)
7. Laser capture microdissection (LCM) system for precise isolation
8. Spectrophotometer or fluorometer for DNA quantification
9. PCR reagents and primers for fungal DNA amplification
10. Sequencing or gel electrophoresis equipment for analysis
11. Bioinformatics tools for data analysis

2.8 Laser Capture Microdissection (LCM)

1. Healthy plant tissues (e.g., leaves, roots, stems)
2. Liquid nitrogen or − 80 °C freezer for tissue preservation
3. Cryostat or microtome for sectioning
4. Adhesive-coated slides for mounting tissue sections
5. Histological stains or fluorescent dyes (optional)
6. Laser capture microdissection (LCM) system for isolation
7. DNA extraction kit for small quantities
8. Spectrophotometer or fluorometer for DNA quantification
9. PCR reagents and primers for fungal DNA amplification
10. Gel electrophoresis or sequencing equipment for analysis
11. Bioinformatics tools for data analysis

3 Methods

3.1 Surface Sterilization

Surface sterilization is a crucial first step in isolating fungal endosymbionts from host tissues, ensuring that only internal fungi are cultured. The procedure involves washing plant or animal tissues thoroughly with sterile water to remove any loose debris. This is followed by treating the tissues with a series of disinfectants, typically including ethanol (70%) and sodium hypochlorite (1–2%), for a few minutes to kill any external microbes. After disinfection, the tissues are rinsed multiple times with sterile water to remove any residual chemicals. The sterilized tissues are then cut into small sections, which are placed on sterile nutrient agar plates to allow the internal fungi to grow. This method is essential to avoid contamination from surface-associated fungi, ensuring that only true endosymbionts are isolated [13].

Procedure of Isolation

1. Collect healthy plant tissues (e.g., leaves, roots, stems) from the field or greenhouse and transport them to the laboratory in sterile bags or containers to avoid contamination.
2. Rinse the plant tissues under running tap water to remove soil, debris, and other external contaminants. Cut the tissues into smaller segments (e.g., 1–2 cm) using sterile scissors or a scalpel.
3. Immerse the tissue segments in 70% ethanol for 30–60 s to kill surface microbes. This step is crucial to ensure the elimination of non-endophytic microorganisms.
4. Transfer the tissue segments into a 2–5% sodium hypochlorite (NaOCl) solution and soak them for 2–5 minutes, depending on the tissue type and thickness.
5. After the NaOCl treatment, rinse the tissue segments thoroughly in sterile distilled water three to five times to remove any residual chlorine.
6. To further ensure surface sterilization, briefly immerse the tissue segments in 70% ethanol again for 10–30 s.
7. Rinse the tissue segments one final time in sterile distilled water to remove any remaining ethanol.
8. Blot the sterilized tissue segments dry using sterile filter paper or paper towels.
9. Place the sterilized tissue segments onto sterile culture media (e.g., potato dextrose agar) in Petri dishes using sterile forceps.
10. Incubate the Petri dishes at an appropriate temperature (usually 25–28 °C) for 5–7 days, observing them daily for fungal growth.

11. Subculture any emerging fungal colonies onto fresh media to isolate pure cultures for further identification and characterization.
12. Confirm the endophytic nature of the isolated fungi by performing reinoculation into sterile plant tissues or molecular identification techniques.

3.2 Tissue Fragmentation and Plating

Once surface sterilization is complete, the next step is tissue fragmentation and plating. The sterilized plant or animal tissues are fragmented into small pieces, typically using a sterile scalpel or scissors. These fragments are then placed onto nutrient-rich agar plates, such as potato dextrose agar (PDA) or malt extract agar (MEA), which provide the necessary nutrients for fungal growth. The plates are incubated at an appropriate temperature (usually between 20 and 30 °C) for several days to weeks. As fungi grow from the tissue fragments, individual fungal colonies can be observed. These colonies are then subcultured onto fresh plates to obtain pure isolates. This technique is particularly effective for isolating fungi from roots, stems, leaves, and other solid tissues [13].

Procedure of Isolation

1. Clean the workbench with 70% ethanol or another suitable disinfectant. Prepare all necessary tools and materials, such as sterile forceps, scalpels, Petri dishes, and culture media (e.g., potato dextrose agar).
2. Surface sterilize the plant tissues using a standard protocol (e.g., ethanol and sodium hypochlorite treatment) to ensure the elimination of surface contaminants.
3. Using sterile scissors or a scalpel, cut the sterilized plant tissues into small fragments, typically 1–2 mm in size. Sterilize the tools between cuts to avoid cross-contamination.
4. Place the tissue fragments onto the surface of sterile culture media in Petri dishes. Use sterile forceps to gently press the fragments into the media, ensuring good contact without embedding them too deeply.
5. Seal the Petri dishes with Parafilm or another suitable sealing material to prevent contamination. Incubate the Petri dishes at an appropriate temperature (usually 25–28 °C) for 5–7 days.
6. Check the plates daily for signs of fungal growth. Monitor for contamination and discard any plates that show signs of non-endophytic growth.
7. Once fungal colonies emerge from the tissue fragments, use a sterile loop or needle to transfer the colonies onto fresh culture media to obtain pure isolates. Continue to incubate and subculture until pure, uncontaminated fungal colonies are obtained.

8. Store the isolated fungal endosymbionts on suitable media under conditions that promote long-term viability (e.g., cryopreservation or storage in glycerol at −80 °C) for future studies.
9. Record all observations, including the number of colonies, morphology, and any other relevant data, for future reference and identification.

3.3 Serial Dilution and Plating

Serial dilution and plating is a method used to isolate fungal endosymbionts from liquid or finely ground samples. The sample is first suspended in a sterile liquid, such as distilled water or a saline solution. A series of dilutions are made by transferring a small volume of the sample to a new tube containing a sterile solution and repeating this process several times. Each dilution is then spread onto agar plates, and the plates are incubated. As the dilution reduces the density of fungi and other microbes, individual fungal colonies can be more easily isolated. This technique is particularly useful when the sample contains a high concentration of microbes, allowing for the separation of fungal endosymbionts from other organisms [14].

Procedure of Isolation

1. Prepare the work area by cleaning the bench with 70% ethanol or another suitable disinfectant. Gather all necessary materials, including sterile test tubes, pipettes, culture media (e.g., potato dextrose agar), and sterile water or buffer.
2. Surface sterilize the plant tissue samples using an appropriate sterilization protocol (e.g., ethanol and sodium hypochlorite treatment) to eliminate surface contaminants.
3. After sterilization, aseptically cut the plant tissue into small pieces (1–2 mm) using sterile scissors or a scalpel.
4. Transfer the tissue pieces into a sterile mortar or blender, and add a small volume of sterile water or buffer (e.g., 10–20 mL) to homogenize the tissue. Grind or blend the tissue until a fine suspension is obtained.
5. Perform the first dilution by transferring 1 mL of the tissue suspension into a sterile test tube containing 9 mL of sterile water or buffer, creating a 1:10 dilution.
6. Mix the first dilution thoroughly by vortexing or shaking. Perform a serial dilution by transferring 1 mL from the first dilution into a second test tube containing 9 mL of sterile water or buffer, creating a 1:100 dilution.
7. Continue the serial dilution process to achieve the desired dilution series (e.g., 10^{-1}, 10^{-2}, 10^{-3}), depending on the expected fungal load in the tissue.

8. Plate 100 μL of each dilution onto sterile culture media in Petri dishes. Use a sterile spreader or glass rod to evenly distribute the suspension across the surface of the agar.
9. Seal the Petri dishes with Parafilm or another suitable sealing material to prevent contamination. Incubate the dishes at an appropriate temperature (usually 25–28 °C) for 5–7 days.
10. Check the plates daily for fungal colony growth. Select plates with well-separated colonies for further analysis, noting the dilution that provided optimal colony separation.
11. Using a sterile loop or needle, transfer isolated fungal colonies from the plates to fresh culture media to obtain pure isolates. Continue subculturing until pure, uncontaminated fungal colonies are obtained.
12. Store the isolated fungal endophytes on suitable media under conditions that promote long-term viability (e.g., cryopreservation or storage in glycerol at −80 °C) for future studies.
13. Record all observations, including the number of colonies per dilution, colony morphology, and any other relevant data, for future reference and identification.

3.4 Single Spore Isolation

Single spore isolation is a precise technique used to obtain pure fungal cultures from mixed colonies. The process involves harvesting spores from a fungal colony using a fine needle or loop under a microscope. These spores are then diluted in a sterile solution and spread onto an agar plate. Alternatively, the spores can be directly picked and transferred to an agar medium. After incubation, individual colonies derived from single spores are observed and subcultured to obtain pure isolates. This method is especially important when dealing with fungi that reproduce via spores, as it ensures that only one fungal strain is isolated [15].

Procure of Isolation

1. Collect plant tissue suspected of harboring fungal endosymbionts.
2. Surface sterilize the tissue by sequentially washing it in 70% ethanol for 1 minute, followed by 1–2% sodium hypochlorite for 3–5 minutes, and rinse thoroughly with sterile distilled water three to four times.
3. Excise small sections of the sterilized tissue using a sterile scalpel or forceps and place them on a sterile Petri dish.
4. Transfer a small portion of the excised tissue onto a sterile microscope slide.
5. Gently crush the tissue with a sterile scalpel or pestle to release the fungal spores.

6. Add a drop of sterile water to the slide and mix gently to create a spore suspension.
7. Using a sterile inoculation loop or needle, transfer a small drop of the spore suspension onto a water agar plate.
8. Place the water agar plate under a dissecting microscope to visualize individual spores.
9. Identify and carefully pick a single spore using a fine needle or an eyelash tool.
10. Transfer the selected spore onto a freshwater agar plate.
11. Incubate the plate at 25–28 °C until the spore germinates and a mycelial colony begins to form.
12. Once the colony has grown sufficiently, transfer a small portion to a new plate containing an appropriate growth medium, such as potato dextrose agar, to establish a pure culture.
13. Continue to subculture as needed to maintain purity and ensure the isolated culture is free from contamination.
14. Use the pure culture for subsequent morphological, molecular identification, or functional characterization studies.

3.5 Symbiotic Culture Techniques

Symbiotic culture techniques involve maintaining the natural symbiotic relationship between the host and the fungal endosymbiont during the isolation process. In this method, host tissues (such as plant roots or insect guts) are cultured together with fungi on specialized media that mimic the host's natural environment. This approach can help in isolating fungi that are highly dependent on their host for growth. The host tissues are placed on or near nutrient media that supports both the host and the fungi, allowing the fungi to grow in a way that maintains their symbiotic relationship. This method is particularly useful for isolating endosymbionts that are difficult to culture independently [16].

Procedure of Isolation

1. Clean the workbench with 70% ethanol or another suitable disinfectant. Gather necessary materials, including sterile Petri dishes, culture media (e.g., water agar), sterile water, fine needles, a dissecting microscope, and sterile pipettes.
2. Subculture the fungal endosymbionts on a suitable growth medium (e.g., potato dextrose agar) and incubate under appropriate conditions (usually 25–28 °C) until sporulation occurs. Sporulation may take several days to a week, depending on the fungal species.
3. Using a sterile loop or needle, gently scrape the spores from the sporulating culture into a small volume of sterile water or buffer (e.g., 1–2 mL) in a sterile test tube. Vortex or shake the tube gently to create a uniform spore suspension.

4. Perform a serial dilution of the spore suspension to reduce the spore concentration, making it easier to isolate single spores. For example, transfer 1 mL of the spore suspension into 9 mL of sterile water or buffer, creating a 1:10 dilution. Repeat as necessary to achieve the desired dilution.
5. Plate 100 μL of the diluted spore suspension onto the surface of sterile water agar in a Petri dish. Spread the suspension evenly across the agar surface using a sterile spreader or glass rod. Water agar is used to limit extensive hyphal growth, which facilitates the isolation of individual spores.
6. Seal the Petri dishes with Parafilm or another suitable sealing material to prevent contamination. Incubate the plates at an appropriate temperature (usually 25–28 °C) for 12–24 hours, allowing the spores to settle and adhere to the agar surface without germinating.
7. Under a dissecting microscope, locate well-separated single spores on the agar surface. Using a fine sterile needle or micromanipulator, carefully pick individual spores and transfer them to fresh Petri dishes containing sterile culture media (e.g., potato dextrose agar).
8. Incubate the dishes containing the single spores at an appropriate temperature (usually 25–28 °C) until the spores germinate and form fungal colonies. This process may take several days, depending on the fungal species.
9. Once colonies have formed from the single spores, transfer them to fresh culture media using a sterile loop or needle to obtain pure fungal cultures. Continue subculturing until pure, uncontaminated fungal colonies are obtained.
10. Store the isolated fungal endosymbionts on suitable media under conditions that promote long-term viability (e.g., cryopreservation or storage in glycerol at −80 °C) for future studies [16].
11. Record all observations, including spore morphology, colony formation, and any other relevant data, for future reference and identification.

3.6 Culture-Independent Methods

Culture-independent methods bypass the need for culturing fungi and instead rely on molecular techniques to identify and study fungal endosymbionts. One common approach is DNA extraction from host tissues followed by PCR amplification of fungal-specific genetic markers, such as the internal transcribed spacer (ITS) region or 18S rRNA genes. These amplified DNA sequences can then be sequenced and compared to known fungal databases to identify the endosymbionts present. Metagenomics, which involves high-throughput sequencing of all DNA in a sample, can also be used to uncover the diversity of fungal endosymbionts without culturing. These methods are particularly valuable for studying non-culturable fungi and for understanding the entire fungal community within a host [17].

Procedure

1. Collect healthy plant tissues (e.g., leaves, roots, stems) from the field or greenhouse and transport them to the laboratory in sterile containers to avoid contamination.
2. Surface sterilize the plant tissues using a standard protocol, such as a series of washes with ethanol and sodium hypochlorite, followed by rinsing with sterile water, to remove external microorganisms (*see* **Note 18**).
3. Homogenize the surface-sterilized plant tissue using a sterile mortar and pestle or a mechanical tissue homogenizer. Extract total DNA from the homogenized tissue using a DNA extraction kit or protocol designed for plant tissues, ensuring the recovery of DNA from both the plant and the fungal endosymbionts (*see* **Notes 19–25**).
4. Quantify the extracted DNA using a spectrophotometer or fluorometer, and assess the quality by running the DNA on an agarose gel to check for degradation or contamination.
5. Perform PCR amplification using primers targeting fungal-specific DNA regions, such as the internal transcribed spacer (ITS) regions of ribosomal DNA, to selectively amplify fungal DNA from the total DNA extract.
6. Run the PCR products on an agarose gel to verify the successful amplification of the target DNA. If needed, excise the desired bands and purify the PCR products using a gel extraction kit.
7. Clone the purified PCR products into a suitable plasmid vector using a cloning kit, and transform the cloned plasmids into competent bacterial cells. Plate the cells on selective media to isolate individual colonies, pick positive colonies, and sequence the inserted fungal DNA to identify the fungal endosymbionts.
8. Alternatively, prepare the PCR-amplified fungal DNA for next-generation sequencing (NGS) using a library preparation kit, and sequence the DNA using an NGS platform (e.g., Illumina) to identify and quantify fungal endosymbionts based on the ITS or other fungal-specific regions.
9. Process the sequencing data using bioinformatics tools to identify fungal endosymbionts. Compare the sequences to fungal databases (e.g., UNITE, NCBI) to classify and annotate the fungal species present in the sample.
10. Analyze the diversity, abundance, and potential functions of the fungal endosymbionts in the plant tissue. Document the

methodologies used, the species identified, and their potential roles in the plant's biology in a report or publication.

11. Store the extracted DNA and sequencing data under appropriate conditions for future reference. DNA samples can be stored at −20 °C or − 80 °C, and sequencing data should be backed up and archived in secure databases.

3.7 Microscopy-Guided Isolation

Microscopy-guided isolation combines traditional isolation techniques with microscopic examination. In this method, host tissues are first examined under a microscope to locate specific fungal structures, such as hyphae, spores, or fruiting bodies. Once identified, these structures are carefully isolated using fine needles or micromanipulators and transferred to nutrient agar plates for culturing. This technique allows for the precise isolation of specific fungal endosymbionts, especially those that form distinct structures within the host tissues. It is particularly useful in isolating fungi that are present in low abundance or are embedded deep within host tissues [18].

Procedure of Isolation

1. Collect healthy plant tissues (e.g., leaves, roots, stems) from the field or greenhouse and transport them to the laboratory in sterile containers to avoid contamination.
2. Surface sterilize the plant tissues using a standard protocol, such as a series of washes with ethanol and sodium hypochlorite, followed by rinsing with sterile water, to remove external microorganisms.
3. Homogenize the surface-sterilized plant tissue using a sterile mortar and pestle or a mechanical tissue homogenizer. Extract total DNA from the homogenized tissue using a DNA extraction kit or protocol designed for plant tissues, ensuring the recovery of DNA from both the plant and the fungal endosymbionts.
4. Quantify the extracted DNA using a spectrophotometer or fluorometer, and assess the quality by running the DNA on an agarose gel to check for degradation or contamination.
5. Perform PCR amplification using primers targeting fungal-specific DNA regions, such as the internal transcribed spacer (ITS) regions of ribosomal DNA, to selectively amplify fungal DNA from the total DNA extract.
6. Run the PCR products on an agarose gel to verify the successful amplification of the target DNA. If needed, excise the desired bands and purify the PCR products using a gel extraction kit.
7. Clone the purified PCR products into a suitable plasmid vector using a cloning kit, and transform the cloned plasmids into competent bacterial cells. Plate the cells on selective media to

isolate individual colonies, pick positive colonies, and sequence the inserted fungal DNA to identify the fungal endosymbionts.

8. Alternatively, prepare the PCR-amplified fungal DNA for next-generation sequencing (NGS) using a library preparation kit, and sequence the DNA using an NGS platform (e.g., Illumina) to identify and quantify fungal endosymbionts based on the ITS or other fungal-specific regions.
9. Process the sequencing data using bioinformatics tools to identify fungal endosymbionts. Compare the sequences to fungal databases (e.g., UNITE, NCBI) to classify and annotate the fungal species present in the sample.
10. Analyze the diversity, abundance, and potential functions of the fungal endosymbionts in the plant tissue. Document the methodologies used, the species identified, and their potential roles in the plant's biology in a report or publication.
11. Store the extracted DNA and sequencing data under appropriate conditions for future reference. DNA samples can be stored at −20 °C or − 80 °C, and sequencing data should be backed up and archived in secure databases.

3.8 Laser Capture Microdissection (LCM)

Laser capture microdissection (LCM) is a high-precision technique that enables the isolation of specific fungal cells or tissues from host organisms. Using a laser under a microscope, researchers can cut and capture targeted areas of the host tissue that contain fungal endosymbionts (20). These microdissected samples can then be transferred to culture media or processed for molecular analysis. LCM is particularly valuable when fungal endosymbionts are localized in specific regions of the host tissue or when they are present in very small amounts. This method allows for the isolation of fungal endosymbionts with minimal contamination from surrounding tissues.

Procedure

1. Collect and immediately freeze healthy plant tissues (e.g., leaves, roots, stems) in liquid nitrogen or store them at −80 °C to maintain cellular integrity.
2. Prepare thin sections (5–10 μm) (*see* **Note 25**) of the frozen plant tissues using a cryostat or a microtome. Mount these sections onto specialized adhesive-coated slides compatible with LCM (*see* **Note 26**).
3. Stain the tissue sections if required, using histological stains or fluorescent dyes to enhance the visibility of fungal endosymbionts and plant cells (*see* **Note 27**).
4. Allow the stained tissue sections to air-dry or follow the manufacturer's instructions for additional processing needed for your specific LCM system (*see* **Note 28**).

5. Place the prepared slides into the LCM system. Use the microscope to identify areas of interest containing fungal endosymbionts (*see* **Note 29**).
6. Employ the laser within the LCM system to precisely cut around the regions of interest containing fungal endosymbionts. Capture these regions onto a collection film or cap.
7. Extract the captured material, which includes the fungal endosymbionts and adjacent plant tissue, from the collection film or cap according to the manufacturer's protocol (*see* **Notes 30–32**).
8. Perform DNA extraction from the isolated material using an appropriate DNA extraction kit or protocol suited for small quantities of material.
9. Quantify and assess the quality of the extracted DNA with a spectrophotometer or fluorometer, and verify DNA integrity by running it on an agarose gel.
10. Amplify fungal-specific DNA regions (e.g., internal transcribed spacer (ITS) regions) using PCR with primers designed for fungal identification.
11. Analyze the PCR products through gel electrophoresis or sequencing. Compare the sequences to fungal databases for identification and classification.
12. Document the procedure, including the locations of the microdissected areas and characteristics of the isolated fungi. Compile the findings into a report or publication.
13. Store the extracted DNA and any remaining tissue samples properly for future analysis, and ensure all data is securely archived.

4 Notes

1. When using 70% ethanol for surface disinfection (*see* **Note 1**), it is important to immerse the tissue completely to ensure even coverage. A short dip (about 30 s) is typically sufficient, followed by a rinse in sterile water to remove any ethanol residue that could affect subsequent steps.
2. Sodium hypochlorite (NaOCl) solution should be freshly prepared or stored appropriately to maintain its effectiveness. The concentration and duration of exposure can vary depending on the tissue type; softer tissues may require lower concentrations or shorter exposure times to prevent damage.
3. Sterile filter paper or paper towels should be used for drying tissues after disinfection to avoid contamination. Pat the tissues gently to remove excess liquid without damaging them.

4. Sterile conditions must be maintained throughout the process. All tools (forceps, scissors, scalpel) should be sterilized before use, and tissues should be handled in a laminar flow hood or another sterile environment to minimize the risk of contamination.
5. When cutting tissue for plating, it is crucial to ensure that the tools are sterile and that the cutting surface is also disinfected. Repeated sterilization between samples helps prevent cross-contamination.
6. The tissue fragments should be small enough to promote even contact with the culture medium but large enough to include sufficient microbial content for growth. Generally, fragments of about 5–10 mm in size are effective.
7. Label the Petri dishes clearly with details such as the sample type, date, and medium type. This information is crucial for tracking and analyzing results.
8. Seal the Petri dishes with Parafilm or another sealing material to prevent contamination from the environment while allowing gas exchange. This is particularly important if incubating for extended periods.
9. Serial dilutions should be performed using sterile techniques to avoid introducing contaminants into the sample. Start with a high dilution factor if the sample is expected to have a high microbial load to ensure isolated colonies.
10. When homogenizing tissue, ensure that the equipment (mortar, blender) is thoroughly sterilized between samples to prevent cross-contamination.
11. Plating should be done as quickly as possible after dilution to minimize the loss of viability of the microorganisms. Spread the diluted sample evenly across the plate using a sterile spreader or glass rod to ensure uniform colony distribution.
12. After plating, incubate the plates at the appropriate temperature (usually 25–28 °C) to allow for optimal microbial growth. Regularly check for contamination and colony morphology changes.
13. Sterility is paramount when isolating single spores. Work under a laminar flow hood and ensure all tools, including the loop or needle, are sterile before each use.
14. Visualization under a dissecting microscope is critical for accurately picking single spores. Adjust the magnification as needed to ensure clear visualization of the spores.
15. A gentle touch is required when transferring spores to avoid damaging them. A sterile water or buffer solution can be used

to create a spore suspension, which can then be streaked onto an agar plate for isolation.

16. It is important to incubate the plates under optimal conditions and regularly monitor for spore germination and growth. Subculturing may be necessary to obtain pure cultures from the germinated spores.
17. Host tissues should be handled with care to maintain their viability for symbiotic culture. Work quickly and under sterile conditions to transfer tissues to culture media.
18. Ensure complete removal of sterilizing agents by thorough rinsing, as residual chemicals can degrade DNA.
19. Incubation conditions should mimic the natural environment of the host and fungi as closely as possible to promote successful symbiotic interactions. Monitoring the cultures regularly is key to observing the development of symbiosis.
20. When transferring cultures, it is important to avoid introducing contaminants. Use sterile pipettes and other tools, and work within a sterile environment to ensure the purity of the cultures.
21. Homogenization of tissue should be thorough to ensure that cells are lysed and DNA is released. However, over-homogenization can shear DNA, so a balance is required.
22. The choice of a DNA extraction kit can significantly impact the quality and yield of DNA. Select a kit that is specifically designed for the type of tissue you are working with to maximize DNA recovery.
23. PCR amplification requires careful optimization of reagents and conditions to ensure specific and efficient amplification of fungal DNA. Use appropriate controls to monitor for contamination and amplification efficiency.
24. Bioinformatics tools are crucial for analyzing sequencing data. Ensure that the software and databases used are up-to-date and appropriate for the type of fungi being studied.
25. Tissue sectioning should be done carefully to avoid damaging the structures of interest. Use a cryostat or microtome set to the appropriate temperature and thickness for your tissue type.
26. Adhesive-coated slides are important for mounting tissue sections without losing material during subsequent handling. Ensure that the slides are clean and the adhesive is fresh for optimal results.
27. Histological stains or fluorescent dyes can enhance the visibility of specific structures within the tissue. Choose the staining method based on the target structures and the downstream analysis.

28. Laser capture microdissection (LCM) requires precise handling and focus. Ensure that the LCM system is calibrated and that the tissue sections are appropriately prepared for successful isolation of the regions of interest.
29. Tissue preservation is crucial for maintaining the integrity of the sample during sectioning and LCM. Freeze the tissues quickly in liquid nitrogen or store them at −80 °C to prevent degradation.
30. Sectioning should be performed under sterile conditions to avoid introducing contaminants. Ensure the cryostat or microtome is clean and properly maintained to produce high-quality sections.
31. The LCM system should be calibrated regularly to ensure precision in capturing the desired areas. The use of adhesive-coated slides can help in maintaining the integrity of the captured sections during the process.
32. DNA extraction from the captured sections requires careful handling due to the small quantities of material. Use a kit designed for low-input samples to maximize DNA recovery and ensure accurate downstream analysis.

References

1. Valadez-Cano C, Olivares-Hernández R, Espino-Vázquez AN, Partida-Martínez LP (2024) Genome-scale model of *Rhizopus microsporus*: metabolic integration of a fungal holobiont with its bacterial and viral endosymbionts. Environ Microbiol 26:e16551
2. Rosling A, Eshghi Sahraei S, Kalsoom Khan F, Desirò A, Bryson AE, Mondo SJ, Sánchez-García M (2024) Evolutionary history of arbuscular mycorrhizal fungi and genomic signatures of obligate symbiosis. BMC Genomics 25:529
3. Pawlowska TE, Gaspar ML, Lastovetsky OA, Mondo SJ, Real-Ramirez I, Shakya E, Bonfante P (2018) Biology of fungi and their bacterial endosymbionts. Annu Rev Phytopathol 56: 289–309
4. Bianciotto V, Lumini E, Lanfranco L, Minerdi D, Bonfante P, Perotto S (2000) Detection and identification of bacterial endosymbionts in arbuscular mycorrhizal fungi belonging to the family Gigasporaceae. Appl Environ Microbiol 66:4503–4509
5. Suryanarayanan TS (2011) Fungal endosymbionts of seaweeds. In: Biology of marine fungi. Springer, Berlin/Heidelberg, pp 53–69
6. Lecomte J, St-Arnaud M, Hijri M (2011) Isolation and identification of soil bacteria growing at the expense of arbuscular mycorrhizal fungi. FEMS Microbiol Lett 317:43–51
7. Mahulette AS, Alfian A, Kilkoda AK, Lawalata IJ, Marasabessy DA, Tanasale VL, Makaruku MH (2021) Isolation and identification of indigenous arbuscular mycorrhizal fungi (AMF) of forest clove rhizosphere from Maluku, Indonesia. Biodiversitas 22:3728–3736
8. Sikora RA, Pocasangre L, zum Felde A, Niere B, Vu TT, Dababat AA (2008) Mutualistic endophytic fungi and in-planta suppressiveness to plant parasitic nematodes. Biol Control 46:15–23
9. Saikkonen K, Faeth SH, Helander M, Sullivan TJ (1998) Fungal endophytes: a continuum of interactions with host plants. Annu Rev Ecol Syst 29:319–343
10. Varma A, Verma S, Sudha Sahay N, Bütehorn B, Franken P (1999) Piriformospora indica, a cultivable plant-growth-promoting root endophyte with similarities to arbuscular mycorrhizal fungi. Appl Environ Microbiol 65: 2741–2744
11. Harman GE, Howell CR, Viterbo A, Chet I, Lorito M (2004) Trichoderma species—opportunistic, avirulent plant symbionts. Nat Rev Microbiol 2:43–56

12. Rajagopal K, Suryanarayanan TS (2000) Isolation of endophytic fungi from leaves of neem (*Azadirachta indica* A. Juss.). Curr Sci 79: 1375–1378
13. Gautam AK, Kant M, Thakur Y (2013) Isolation of endophytic fungi from *Cannabis sativa* and study of their antifungal potential. Arch Phytopathol Plant Protect 46:627–635
14. Sanders ER (2012) Aseptic laboratory techniques: plating methods. J Vis Exp 63:e3064
15. Choi YW, Hyde KD, Ho WH (1999) Single spore isolation of fungi. Fungal Divers 3:29–38
16. Kumari R, Giang PH, Sachdev M, Garg AP, Varma A (2004) Symbiotic fungi for eco-friendly environment: a perspective. Curr Sci 86:1375–1378
17. Wijayawardene NN, Bahram M, Sánchez-Castro I, Dai DQ, Ariyawansa KG, Jayalal U, Tedersoo L (2021) Current insight into culture-dependent and culture-independent methods in discovering Ascomycetous taxa. Mycol Prog 20:657–669
18. Vicente VA, Angelis DAD, Queiróz-Telles Filho F, Pizzirani-Kleiner AA (2001) Isolation of herpotrichiellacious fungi from the environment. Braz J Microbiol 32:47–51

Chapter 6

Isolation and Characterisations of Endosymbionts from Slime Mold Fruiting Bodies

Jitendra D. Salunkhe, Atharva S. Patil, and Satish V. Patil

Abstract

Slime molds, though morphologically similar to fungi, are protists and exhibit a range of colors including red, brown, yellow, and blue. They are commonly found in moist environments rich in decaying organic matter, such as leaf litter, wood, and agricultural or forest waste. Slime molds are classified into two major types: cellular and acellular. Their life cycle includes two distinct stages: an amoeboid feeding stage, during which they consume bacteria, fungi, and organic debris, and a fruiting stage that resembles a fungal structure. Recently, slime molds have gained importance as a source of new bioactive compounds. Several metabolites, such as naphthoquinones, pyrones, tetramic acid derivatives, and physarochrome, have been reported from several slime molds and their fruiting bodies. While associated endosymbiotic bacteria have been reported, they remain comparatively underexplored. This chapter focuses on the isolation of endosymbionts from slime mold fruiting bodies, with the aim of elucidating their role in the host's life cycle and secondary metabolite production.

Key words Protozoa, Symbiotic association, Acellular slime mold

1 Introduction

Slime molds are a diverse group of eukaryotic organisms, characterized by two distinct forms: a motile amoeboid stage and a reproductive, spore-producing stage. Slime molds belong to the kingdom Protista and possess characteristics that are intermediate between fungi and protists [1]. Although they resemble fungi in appearance and form spore-producing fruiting bodies, slime molds were originally misclassified as fungi until their true protistan nature was recognized. Slime molds exhibit a unique life cycle, existing both as unicellular amoebae and large multicellular aggregates. They are broadly classified into two major types: plasmodial and cellular slime molds. Plasmodial slime molds have a considerable and naked plasmodium stage that moves and feeds as a single large organism. In contrast, cellular slime molds, such as

N. Amaresan and Sapna Chandwani (eds.), *Endosymbionts*, Springer Protocols Handbooks,
https://doi.org/10.1007/978-1-0716-4771-4_6,

Dictyostelium, exist as individual amoeboid cells that aggregate to form a multicellular slug in response to nutrient scarcity [2].

Previous studies have suggested that slime molds are among the oldest and smallest known farmers on Earth. These single-celled organisms that are regarded as "ancient farmers" cultivate bacteria as a food source. Some molds are capable of attracting and retaining bacteria while migrating, subsequently using them to seed new food sources once their current environment becomes depleted. An example of this behavior is seen in *Dictyostelium discoideum* [3, 4]. It has also been reported that certain *Pseudomonas* spp. exhibit endosymbiotic relationships with the fruiting bodies of cellular slime molds and may serve as a bacterial food reserve. However, the precise nature of this interaction remains unclear and may involve additional, yet unexplored, biological roles [5, 6].

This study presents the isolation of a novel endosymbiotic bacterium from acellular slime molds, which, to our knowledge, has not been studied previously. The mechanisms by which these endosymbiotic bacteria associate with slime molds remain unclear, largely due to the limited number of studies focused on these interactions. The relationship between endosymbionts and the fruiting bodies of slime molds is still largely unexplored.

The slime mold and their endosymbionts are increasingly being considered as potential sources of new antimicrobial compounds. Given the global rise in multidrug-resistant pathogens, the discovery of new bioactive metabolites is critically important. Various reports have highlighted the potent bioactivity of slime mold–derived compounds such as pigments, mucous glycolipids, and acyltetramic acids against pathogens, including herpes simplex virus (HSV) and antibiotic-resistant bacteria [7]. In recent years, multiple secondary metabolites, including naphthoquinones, pyrones, tetramic acid derivatives, and physarochrome, have been reported from several slime molds and their fruiting bodies. While a few studies have also reported bioactive metabolites from slime mold endosymbionts, this area remains relatively underexplored. Dembitsky et al. (2005) reported that myxomycetes are able to produce at least 100 bioactive natural products. These metabolites may serve as valuable sources for future antiviral and antibacterial drug development. Although there are a few reports on mold-associated endosymbionts, their specific roles in secondary metabolite production and in regulating the slime mold life cycle remain poorly understood [7]. This study focuses on the isolation and characterization of fluorescent *Pseudomonas* spp. from common slime molds, such as *Physarum* spp. The findings may support future investigations into the ecological roles, interactions, and potential biotechnological applications of these endosymbiotic bacteria.

2 Materials

1. Slime mold fruiting body sample.
2. 10× magnifying lens.
3. Sterile sample bags.
4. Marker.
5. Sterile saline solution (0.85% NaCl in 100 mL distilled water).
6. Cetrimide agar medium (peptone 20 g/L, potassium sulphate 10 g/L, magnesium chloride 1.4 g/L, cetrimide 0.3 g/L, agar 30 g/L, final pH 7.0 ± 0.2 at 25 °C).
7. Sterile Petri plates.
8. Sterile forceps.
9. Grease-free microscope slides.
10. Coverslips.
11. Sterile glass rod.
12. Nichrome wire loop.
13. Autoclave.
14. UV chamber.
15. Incubator.
16. Compound microscope 100x.
17. Staining reagents.
18. 70% ethanol.

3 Methods

3.1 Collection of Slime Mold Fruiting Bodies from Decaying Wood and Leaves

1. Collect slime mold fruiting body samples using sterile plastic bags, gloves, and sterilized tools such as forceps or tweezers.
2. Identify fruiting bodies on moist, decaying wood or leaves, then gently pull the fruiting bodies along with a small portion of the attached substrate, and place them into the sterile bag.
3. Seal the bag tightly and label it properly with relevant information, including the type of substrate (wood or tree material), place, and date of collection (e.g., *Cassia tora* leaves, *Acacia nilotica* bark, *Terminalia bellirica* wood, KBC North Maharashtra University campus, Jalgaon/August 10, 2024).

   Store the sample in a cool place until it is transported to the laboratory for further processing.

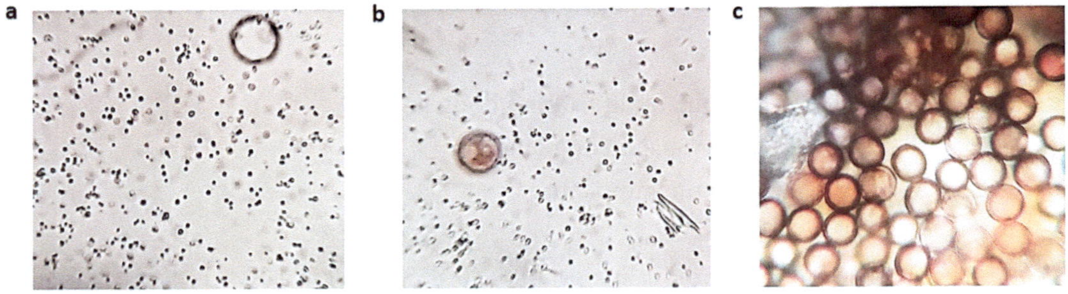

Fig. 1 (**a** & **b**) Endosymbiotic bacteria observed in slime mold fruiting body preparation. c Spores of slime mold observed at 100× magnification https:/youtube.com/shorts/QA-Z8jS_yZs

3.2 Study of Endosymbionts in Slime Mold Fruiting Bodies under Microscopic Examination

1. Observe the collected slime mold fruiting bodies and note their morphological features and characteristic colors. Capture photographs as needed.
2. Separate the fruiting bodies, which are generally hydrophobic, and rinse them with 70% ethanol to remove surface contaminants.
3. Transfer the fruiting bodies into sterile saline using sterile pointed forceps. Using sterile pointed needles or a sterile glass rod, crush the material in a sterile microcentrifuge tube (*see* **Note 1**).
4. Place approximately 200 μL of the prepared sample onto a grease-free glass slide and carefully cover it with a coverslip.
5. Add one drop of oil immersion onto the coverslip for clear and good resolution under the microscope.
6. Examine the slide under a microscope at 100× magnification to observe endosymbiotic bacteria and, in some species, visible spores (*see* Fig. 1).

3.3 Isolation of Endosymbiont on Culture Medium

1. Prepare sterile nutrient, cetrimide, or MacConkey agar plates by pouring the sterile medium aseptically into sterile Petri plates and allow to solidify.
2. Add the separated and surface-sterilized fruiting bodies in sterile saline or buffer solution.
3. Crush the fruiting bodies in a sterile tube using sterile pointed needles or a sterile glass rod.
4. Streak or spread the homogenized sample onto nutrient, cetrimide, MacConkey agar using a sterile nichrome loop.
5. Incubate the plates at 37 °C for a minimum of 24–48 h.
6. Observe the bacterial colonies obtained and record their morphological characteristics (*see* **Notes 2** and **3**) and identify further by molecular techniques (*see* **Note 4**).

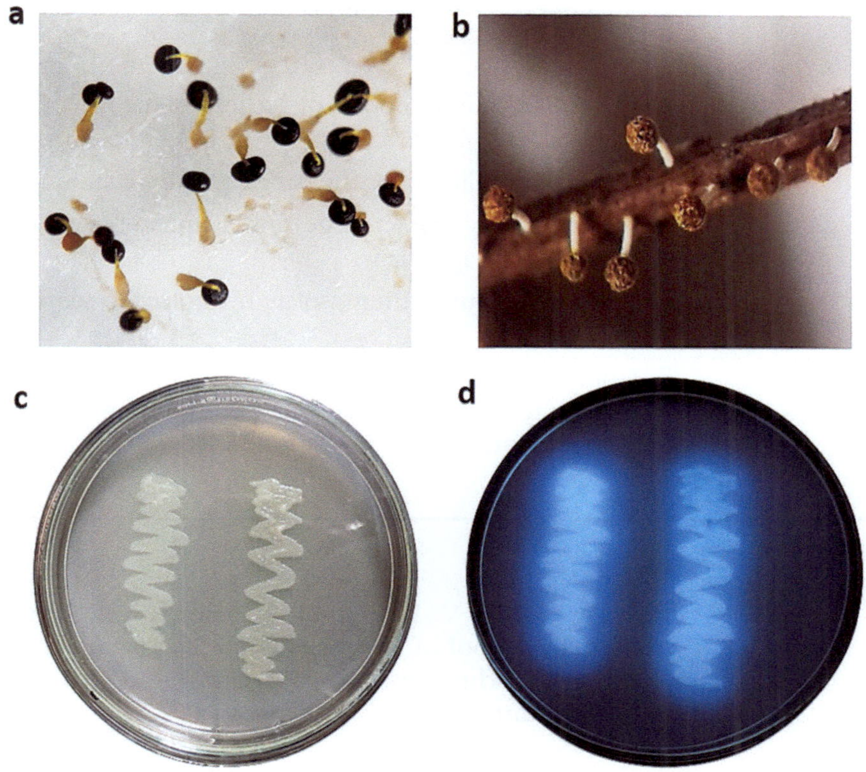

Fig. 2 (**a** & **b**) *Physarum* sp. (slime mold) fruiting bodies (10× magnification). (**c**) Endosymbiont *Pseudomonas* spp. growth on cetrimide agar medium. (**d**) *Pseudomonas* spp. growth under short UV light shows fluorescence

7. In the majority of samples, *Pseudomonas* spp. are commonly isolated. These typically appear as circular colonies (1–4 mm in diameter), yellowish or colorless, and transparent on cetrimide agar. The isolates are Gram-negative and motile. Under UV light, colonies often exhibit greenish, yellowish, or fluorescent pigmentation (*see* **Note 5**, Fig. 2).

3.4 Staining of Isolated Endosymbiont Bacteria

1. Prepare a bacterial suspension of the endosymbiont in sterile saline and make a thin smear on a microscope slide. Heat fix the smear.
2. Stain the smear with crystal violet for 1 min.
3. Apply Gram's iodine solution for 1 min to fix the dye.
4. Decolorize with 95% ethanol or acetone until the runoff appears clear.
5. Counterstain with safranin for 1 min.
6. Rinse the slide with water, blot it dry, and examine it under the microscope. Gram-negative bacteria appear pink under the microscope.

4 Notes

1. Crush the fruiting bodies of acellular slime molds in 2 mL of sterile saline.
2. Observe motility using the hanging drop technique.
3. Record colony characteristics and maintain the endosymbionts on a nutrient agar slant.
4. Identify the pure culture isolate by molecular sequencing, i.e., 16S rRNA.
5. Confirm the identity of *Pseudomonas* spp., if necessary, through further testing, and retain the isolate on an agar slant or as glycerol stocks.

Acknowledgement

Prof. Dr. Satish V Patil kindly acknowledges **Prof. Robert J. Mitchell, School of Life Sciences**, Ulsan National Institute of Science and Technology, Ulsan, South Korea, for his constant support and inspiration.

References

1. Baldauf SL, Doolittle WF (1997) Origin and evolution of the slime molds (Mycetozoa). Proc Natl Acad Sci 94(22):12007–12012
2. Bonner JT, Lamont DS (2005) Behavior of cellular slime molds in the soil. Mycologia 97:178–184
3. Brock DA, Douglas TE, Queller DC, Strassmann JE (2011) Primitive agriculture in a social amoeba. Nature 469:393–396
4. Boomsma JJ (2011) Farming writ small: social slime moulds graze on bacteria, but save some for transmission in their spores. Strains practising this primitive form of farming coexist with non-farmer strains in an intriguing cost-benefit equilibrium. Nature 469:308–310
5. Raper KB (1951) Isolation, cultivation, and conservation of simple slime molds. Q Rev Biol 26: 169–190
6. Bozzone DM (1997) Feeding behaviors in cellular slime molds: a microbial system to study competition. ABT 59:565–572
7. Dembitsky VM, Rezanka T, Spízek J, Hanus LO (2005) Secondary metabolites of slime molds (myxomycetes). Phytochem 66:747–769

Chapter 7

Isolation of Endosymbiotic Actinobacteria

Lakshmaiah Manjunatha, Kotipalli Satya Naga Manideep, Kuracha Tulasi Priya, Gandrajapally Shivareddy Madhu, and Venkataravanappa Venkataravanappa

Abstract

Endosymbiotic organisms live within another organism, and this relationship can be obligated, where one or both organisms depend on each other for survival. Actinobacteria are recognized for producing primary and secondary metabolites with broad applications in industry and medicine. They are also found in marine sponges and serve as endosymbionts in genera such as *Mycobacterium*, *Streptomyces*, and *Salinispora*. These bacteria contribute to the production of antibiotics and other bioactive compounds. Actinomycetes, including *Streptomyces*, exhibit antibacterial, antifungal, antibiofilm, and anticancer properties. The taxonomy of these bacteria, initially based on morphological traits, has shifted to a molecular approach using 16S rRNA sequences for accurate phylogenetic relationships. Agriculture acts as a biocontrol agent by producing enzymes and antibiotics that protect plants from pathogens, reducing pesticide use. Although their isolation can be challenging due to slow growth, endosymbiotic actinobacteria still offer great promise for sustainable agriculture and biotechnology.

Key words Actinobacteria, Biocontrol, Endosymbiont, Metabolites, Sustainable

1 Introduction

An endosymbiont is an organism that lives inside another, sometimes in a mutually dependent relationship. Actinobacteria, known for producing valuable metabolites and industrial enzymes, are key members of sponge-associated microbial communities. Endosymbiotic associations are common among several actinobacterial genera, notably *Streptomyces*, which produce a wide range of antibiotics such as macrolides, β-lactams, and aminoglycosides [1]. Actinomycetes exhibit diverse biomedical properties, including antibacterial [2], antibiofilm [3], antifungal [4], and anticancer [5] activities. Initially classified by sporulation and morphology, these traits were later found unreliable for phylogenetic analysis. A polyphasic taxonomy using molecular and biochemical tools and 16S

N. Amaresan and Sapna Chandwani (eds.), *Endosymbionts*, Springer Protocols Handbooks,
https://doi.org/10.1007/978-1-0716-4771-4_7,

rRNA sequencing has improved the understanding of *Streptomyces* systematics [6]. *Streptomyces*, commonly isolated from plant tissues [7], are aerobic, gram-positive, filamentous actinobacteria with broad ecological and biotechnological importance.

1.1 Applications of Endosymbiotic Actinobacteria in Agriculture and Other Sectors

Actinobacteria produce key secondary metabolites, including antibiotics, cancer enzyme inhibitors, and immune modulators. They can degrade pollutants like hydrocarbons and pesticides, making them useful in environmental and industrial applications. In agriculture, they suppress plant pathogens, promote growth, and produce antifungal and antibacterial compounds. Commercial strains, such as *Streptomyces griseoviridis*, help control crop diseases caused by *Fusarium* and *Alternaria*. Their diverse metabolic traits are vital in plant biotechnology and biocontrol [8–11] (Fig. 1). Actinobacteria support biocontrol through antibiosis, siderophore production, and induced systemic resistance. *Streptomyces* species produce antibiotics and enzymes that inhibit pathogens and degrade fungal cell walls. Their siderophores limit pathogen access to iron, especially in deficient soils. Activating plant defense pathways reduces reliance on chemical pesticides, making them valuable for sustainable, eco-friendly agriculture [12].

1.2 Isolation of Endosymbiotic Actinobacteria

To isolate endosymbiotic actinobacteria, collect and homogenize samples, filter to extract microbes, and inoculate them onto selective media for incubation (Fig. 2). Isolate and store colonies for further study. Due to their slow growth and specific needs, they use specialized media and pre-treatment. Water samples are filtered to concentrate microbes before plating, and actinobacteria from animal intestines can be isolated using aseptic collection and selective culturing [8, 13].

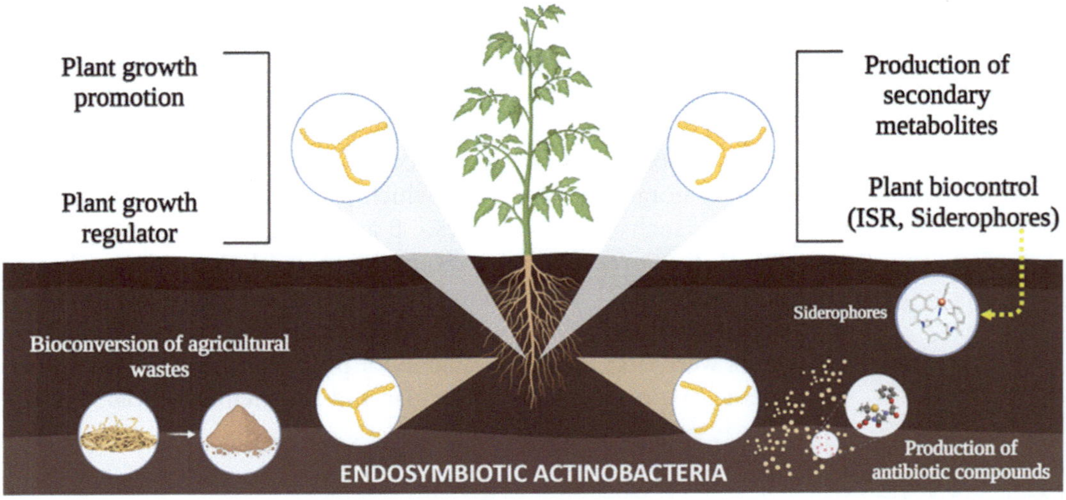

Fig. 1 Sources and application of endosymbiotic actinobacteria

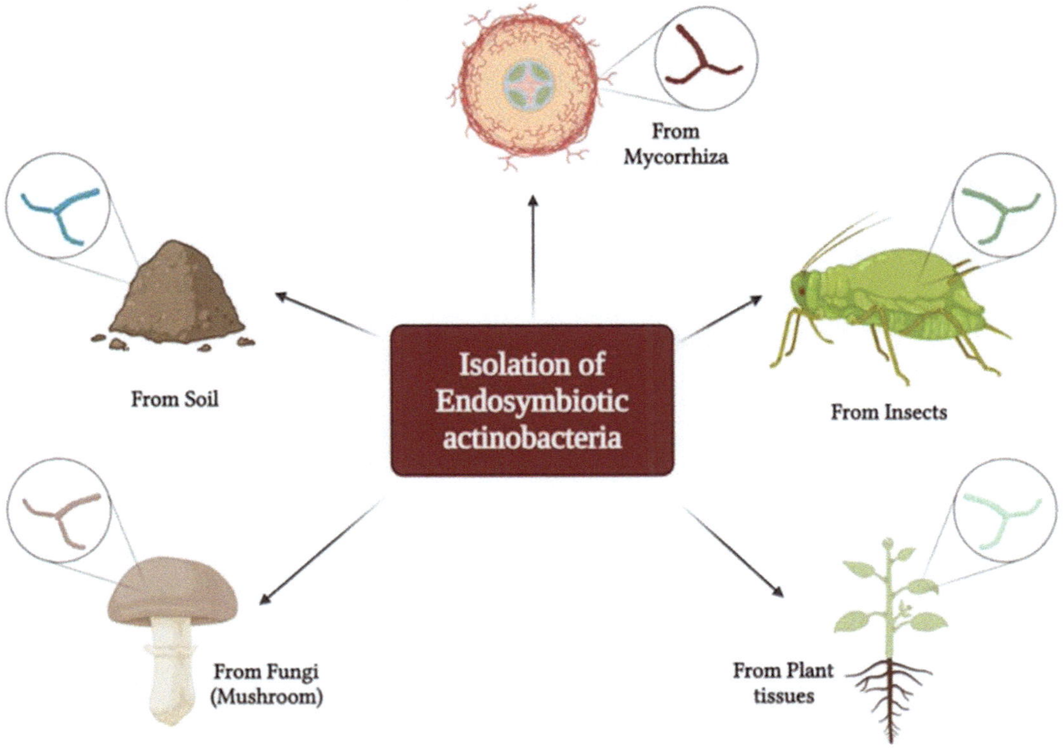

Fig. 2 Isolation sources of endosymbiotic actinobacteria

2 Materials

Prepare all solutions with ultrapure water and analytical-grade reagents to ensure accuracy. Store reagents at room temperature unless specified. Follow all waste disposal regulations carefully, and maintain proper labeling and documentation for safety and compliance.

2.1 Isolation of Actinobacteria from Soil [14]

1. Petri plates.
2. Arginine-glycerol agar.
3. Yeast extract–malt extract agar (ISP2 agar).

2.2 Isolation of Actinobacteria from Soil [15, 16]

1. Starch casein agar (SCA).
2. Actinomycete isolation agar.
3. Glycerol–asparagine agar.
4. Gram staining kit.

5. DNSA extraction kit (the DNSA (2,3-dihydroxy-1-naphthyl-amine) extraction kit is used for the quantification of reducing sugars in biological samples).
6. 16S rRNA gene-specific primers.

2.3 Isolation of Actinobacteria from Plant Tissue [17, 18]

1. Yeast extract–glucose calcium carbonate (YGC).
2. 16S rRNA gene primers (9F: 5′-GAGTTTGATCCTGGCTCAG-3′ and 1510R: 5′-GGCTACCTTGTTACGA-3′) [19].
3. Air dryer.
4. Sonicator.
5. Tap water–yeast extract agar (TWYE).
6. Humic acid–vitamin B agar (HV).
7. Flour–yeast extract–sucrose–casein hydrolysate agar.
8. Yeast extract–casein hydrolysate Dextrose agar (YECD) medium.
9. Ethanol.
10. Sodium hypochlorite.
11. Benomyl.

2.4 Isolation of Actinobacteria from Insects

1. Chitin agar.
2. 0.1% sterile Tween-20 PBS buffer solution.
3. 1.5-mL Eppendorf tubes.
4. PBS solution.
5. Vortex.
6. Yeast–malt extract agar (YMEA).

2.5 Isolation of Actinobacteria from Arbuscular Mycorrhizal Fungi (AMF) Spores [20]

1. Chloramine T trihydrate.
2. Gentamycin sulfate.
3. Streptomycin sulfate.
4. Sterile distilled water.

2.6 Isolation of Endosymbiotic Actinobacteria from Truffles [8]

1. 70% Ethanol.
2. 0.9% sodium hypochlorite.
3. Sterile distilled water.
4. 10% $NaHCO_3$.
5. Chitin–vitamin agar.
6. ISP2 agar: (International Streptomyces Project 2) Agar is a medium commonly used for isolating and cultivating

Streptomyces species and promoting the formation of their characteristic aerial mycelium and spores.

7. Colloidal chitin medium.
8. CAS medium (CAS (chrome azurol S) medium is a specialized medium used primarily for detecting siderophore production by microorganisms. The CAS assay is based on a color change from blue to orange or yellow as siderophores remove iron from the CAS–iron dye complex).
9. Salkowski reagent: It is a chemical solution used primarily to detect the production of indole-3-acetic acid (IAA), and a reagent reacts with IAA to produce a characteristic pink or red color, indicating the presence of this phytohormone.
10. 16S rRNA gene primers.

2.7 Preparation and Screening of Isolates for Antibacterial and Antifungal Activity of Actinobacteria against Plant Diseases [21]

1. Tryptic soy agar.
2. Ethyl acetate or methanol.
3. Sterile Erlenmeyer flasks.
4. Rotary evaporator.
5. Potato dextrose agar.
6. Mueller–Hinton agar.
7. Ethyl acetate or methanol.

3 Methods

Carry out procedures at room temperature unless specified, using calibrated equipment for other conditions. Let reagents and samples reach room temperature before use unless stated otherwise.

3.1 Isolation of Actinobacteria from Soil [14]

1. Collect the soil samples, perform the serial dilutions, and spread the samples on selective media like ISP2 or oatmeal agar. Initially, the isolates will be identified based on morphological traits, such as colony color and shape, and biochemical tests, such as antibiotic production. For precise classification, prefer molecular techniques, particularly 16S rRNA gene sequencing [8, 14].
2. For isolation, suspend 10 g of soil in 100 mL of sterile distilled water and agitate for 30 minutes at 120 rpm.
3. Perform a sixfold dilution and plate 0.1 mL aliquots from each dilution onto arginine–glycerol agar.
4. Incubate the plates for 7 days at room temperature (30 ± 2 °C).
5. Please pick up the pure colonies and transfer them to yeast extract–malt extract agar (ISP2 agar).

3.2 Isolation of Actinobacteria from Soil [15, 16]

1. **Soil Collection and Preparation**
   (a) Collect soil samples from different locations.
   (b) Air-dry soil and sieve it to remove large particles.
2. **Isolation of Actinobacteria**
   (a) Perform serial dilutions of the soil samples.
   (b) Spread the diluted samples onto isolation media such as starch casein agar (SCA), actinomycete isolation agar, or glycerol–asparagine agar.
   (c) Add antifungal agents like cycloheximide to suppress fungal growth.
   (d) Incubate the plates at 28–30 °C for 7 to 14 days.
3. **Morphological Identification**
   (a) Observe colonies for typical actinobacteria characteristics: powdery appearance, earthy smell, and pigment production.
   (b) Use gram staining and microscopy to confirm the presence of filamentous gram-positive bacteria.
4. **Screening for Antimicrobial Activity**
   (a) Subculture-isolated actinobacteria onto fresh media.
   (b) Test the antimicrobial potential using the agar well diffusion method.
   (c) Prepare crude extracts from actinobacteria and evaluate against pathogens like *Staphylococcus aureus*, *Escherichia coli*, and *Candida albicans*.
5. **Molecular Identification**
   (a) Extract genomic DNA from selected isolates.
   (b) Amplify the 16S rRNA gene using PCR.
   (c) Sequence the amplified genes and compare them with known sequences in databases like EzBioCloud for taxonomic identification.

3.3 Isolation of Actinobacteria from Plant Tissue [17]

1. Isolate the endophytic actinobacteria from plant tissues (roots, leaves, stems). For isolation, surfaces sterilize the plant tissues and culture them on selective media like yeast extract–glucose calcium carbonate (YECD).
2. Identify the isolates based on morphological and cultural traits of isolates on ISP media and via molecular techniques such as 16S rRNA gene sequencing for precise classification [21, 22].
3. The plant samples were air-dried for 48 hours before being thoroughly washed to remove all soil from the root mass.

4. The washing includes a sonication step to dislodge soil and organic matter from the roots.
5. Subject the excised roots to a three-step surface sterilization procedure:
   - A 60-second wash in 99% ethanol.
   - A 6-minute wash in 3.125% NaOCl.
   - A 30-second wash in 99% ethanol.
   - A final rinse in sterile reverse osmosis–treated (RO) water.
6. Aseptically, the surface-sterilized roots are sectioned into 1 cm fragments and distributed onto isolation media. They are then allowed to incubate at 27 °C for up to 4 weeks.
7. Several isolation media can be used for the isolation of endophytic actinobacteria, and few methods involve.
   - Tap water–yeast extract agar (TWYE).
   - Humic acid–vitamin B agar (HV).
   - Flour–yeast extract–sucrose–casein hydrolysate agar or yeast extract–casein hydrolysate agar (YECD).
8. Plate all the plant fragments onto one of the prescribed media, and add benomyl (DuPont) at 50 μg/mL to control (endophytic) fungal growth.
9. Separation of endophytes from plant tissue: Grind the plant material and subject it to enzymes that break down plant cell walls. Separate the bacterial pellet by differential centrifugation, dilution, and plating onto isolation media.

3.4 Isolation of Actinobacteria from Insects

1. Isolate actinobacteria from the guts of insects (like termites). First, surface sterilize the insect body parts, then culture them onto selective media. Identify the isolates using morphological assessments and molecular methods such as PCR amplification of the 16S rRNA gene [14].
2. Use chitin agar as the selective medium as a culture-dependent method for isolating actinobacteria [23].
3. In the case of small insects, such as *C. longispina*, *T. ramulorum*, *P. bicornis*, *P. fiorii*, and *P. caeciliae*, samples should consist of five adults, five immature forms (larvae and pupae), and five nest fragments from each colony.
4. Place these samples separately into 0.5 mL of 0.1% sterile Tween-20 PBS buffer solution in sterilized 1.5-mL Eppendorf tubes.
5. Medium-sized insects, including *T. angustula*, *Odontomachus* ants, and all wasps, use pools of three adult insects or larvae. For *P. clavata* ants, only adult insects are collected to isolate actinobacteria.

6. Vortex each sample for 15–20 seconds. Pipette off the Tween-20 PBS solution and discard it, replacing the samples with 0.5 mL of PBS solution.
7. Macerate the samples using a sterile pestle and add another 0.5 mL of PBS. Then, vortex the tubes again for 15 seconds.
8. Plate a 100 μL aliquot from each sample onto chitin agar plates and then incubate at 27 °C for 3 to 4 weeks.
9. Subculture the isolated actinobacteria onto yeast–malt extract agar (YMEA) containing antifungals (nystatin and cycloheximide) to examine their macro- and micromorphology.
10. Perform the gram staining of the isolates before proceeding with DNA extractions of the most promising candidates.
11. To preserve the actinobacteria, prepare a 20% glycerol stock of spores and cells frozen in a homogenized mixture in liquid nitrogen and store them in cryovials at −80 °C.

3.5 Isolation of Actinobacteria from Arbuscular Mycorrhizal Fungi (AMF) Spores [20]

1. Obtain three AMF species: *Claroideoglomus etunicatum* PBT03, *Funneliformis mosseae* RYA08, and *Gigaspora* sp.
2. For isolation of *C. etunicatum* and *F. mosseae*, the rhizosphere soil of teak (*Tectona grandis*) is preferred, while asparagus (*Asparagus officinalis*) is preferred for *Gigaspora* sp.
3. Sterilize 30 spores from each AMF species using chloramine T trihydrate and an antibiotic solution containing gentamycin sulfate and streptomycin sulfate.
4. Wash the spores with sterile distilled water, then wash with water to isolate actinobacteria on specialized agar plates.
5. Incubate the plates at 30 °C for one month. Then, select the actinobacterial colonies, purify them, and store them for further study.

Identification of Actinobacteria

1. Classify actinobacterial strains using stereomicroscopes and light microscopes, referring to *Bergey's Manual of Systematic Bacteriology* and evaluating cell structure, metabolic characteristics, and genetic data.
2. Identify the type of diaminopimelic acid (DAP) in the cell walls through thin-layer chromatography (TLC) to aid in classification and confirm actinobacterial presence based on cell wall composition.
3. Extract genomic DNA from each strain and amplify the 16S rRNA gene using polymerase chain reaction (PCR).
4. Purify the amplified PCR products, sequence them, and analyze the results through the EzBioCloud database for species-level identification.

5. Perform phylogenetic analysis using the neighbor-joining method, and validate the tree topology through bootstrap analysis with 1000 replicates.

Characterization of Plant Growth Promotion Properties

1. Grow actinobacterial strains in broth supplemented with L-tryptophan to induce the production of IAA-like substances [24]. Test auxin production using Salkowski's reagent, where a color shift from pink to red confirms presence. Quantify IAA-like substances via high-performance liquid chromatography (HPLC).
2. Assess siderophore production using the chrome azurol S (CAS) assay, where a visible color change around the colonies indicates positive production [25].
3. Evaluate phosphate solubilization on Pikovskaya (PVK) agar with tricalcium phosphate and potassium solubilization on Aleksandrov agar with potassium feldspar. Determine solubilization efficiency by measuring clear zones around colonies [26].
4. Analyze the production of cell wall–degrading enzymes such as endoglucanase and chitinase on cellulose and colloidal chitin agar. Confirm activity by halo zone formation and express results using the enzyme activity index (EAI) [27].
5. Inoculate strains in peptone water and assess ammonia production by adding Nessler's reagent. A color change confirms the presence of ammonia [28].
6. Test drought tolerance by growing strains on ISP9 agar supplemented with varying concentrations of sorbitol to simulate different water activities (aw) [29].
7. Assess pH tolerance by cultivating strains in ISP2 broth adjusted to a pH range of 3 to 10 [30].
8. Evaluate salinity tolerance by growing strains on ISP2 agar containing NaCl concentrations ranging from 0 to 10% [31].
9. Examine agrochemical tolerance by inoculating strains in ISP2 broth containing different dosages of fungicides, herbicides, and insecticides. Calculate the tolerance index (TI) based on biomass variation between treated and control cultures [32, 33].

3.6 Plant Growth Promotion in Chilli by Actinobacteria and AMF Spores in Greenhouse [20]

1. Surface sterilize chili seeds and germinate them in trays containing sterilized soil.
2. Transplant the seedlings into pots filled with sandy soil amended with 2% coffee waste biochar.
3. Inoculate actinobacterial strains and AMF spores into the soil near the chili seedling root zone.

4. Set up the experiment using a completely randomized design (CRD) with eight treatments under controlled greenhouse conditions for 180 days.
5. Periodically measure plant height, leaf number, stem diameter, branch number, and chlorophyll index.
6. Record fruit yield and assess AMF colonization in the roots.
7. Perform statistical analysis using ANOVA, and determine significant differences ($p \leq 0.05$) using Duncan's multiple range test.

3.7 Isolation of Endosymbiotic Actinobacteria from Truffles [8]

1. Select truffle tubers based on their abundance and adaptability, ensuring no prior actinobacterial isolation studies have been conducted.
2. Harvest *Terfezia leonis* truffles using the Bmark method, identifying cracked or swollen soil as indicators.
3. Place harvested truffles in sterile plastic bags and transport them to the lab for further processing.
4. Wash tubers in running water to remove sand, then surface sterilize by sequential immersion in 70% ethanol (5 min) and 0.9% sodium hypochlorite (20 min), followed by three rinses in sterile distilled water.
5. Soak sterilized tubers in 10% NaHCO for 10 minutes to disrupt tissue for better microbial isolation.
6. Aseptically remove the epidermis and cut the tissue into small cubes for plating on chitin–vitamin agar supplemented with cycloheximide (80 μg/mL) and nalidixic acid (15 μg/mL).
7. Incubate plates at 30 °C for 21 days for actinobacteria growth.
8. Verify sterilization by modifying the method of Cao et al. [34]: Shake sterilized parts in sterile water (250 rpm, 20 min) and plate the wash on chitin–vitamin agar to check for contaminants.
9. Characterize isolated actinobacteria on ISP2, ISP3, and ISP4 media by observing colony traits, pigment production, and spore chain morphology.
10. Identify *Streptomyces* species by analyzing the isomer forms of diaminopimelic acid (DAP) in their cell walls.
11. Evaluate antifungal activity via streak assay against tomato pathogens, inoculating actinobacteria and fungi on ISP2 plates and measuring inhibition zones.
12. Test chitinolytic activity on colloidal chitin medium, recording the diameter of clear zones.
13. Assess siderophore production on CAS agar; the formation of orange or red zones indicates positive results.

14. Determine production of IAA and GA using established methods; detect IAA with Salkowski reagent and GA via extraction and absorbance measurements.
15. Thin-layer chromatography (TLC) is used to identify plant growth regulators by comparing Rf values with standard compounds.
16. For internal colonization studies, surface sterilize and germinate tomato seeds in autoclaved sandy soil, then immerse trimmed roots in actinobacterial suspensions for 24 hours before planting in sterile soil.
17. Prepare microbial suspensions by culturing actinobacteria on ISP2 agar and adjusting concentration to 10^6 CFU/mL.
18. Surface sterilize tomato seeds and coat them with microbial suspension, allowing them to dry before planting.
19. Conduct in vivo growth promotion trials in greenhouse conditions, measuring shoot and root lengths and fresh and dry weights after 6 weeks.
20. Confirm the identity of *Streptomyces* sp. via 16S rRNA gene sequencing, followed by PCR amplification and GenBank submission.
21. Perform all experiments in triplicate, expressing results as mean ± standard deviation.
22. Use ANOVA followed by Fisher's LSD test ($p < 0.05$) for statistical analysis of treatment effects.

3.8 Preparation and Screening of Isolates for Antibacterial and Antifungal Activity of Actinobacteria Against Plant Diseases [21]

3.8.1 Extraction of Bioactive Compounds

1. Culture the isolated actinobacteria strains on tryptic soy agar (TSA) plates at 28 ± 2 °C for 7 days.
2. Harvest each isolate by scraping the surface of the agar and transferring the biomass into sterile Erlenmeyer flasks containing ethyl acetate or methanol as solvents.
3. Shake the flasks at room temperature for 24 hours to extract bioactive compounds.
4. Filter the mixture to separate the biomass from the solvent.
5. Concentrate the filtrate using a rotary evaporator to remove the solvent, which will result in a dry extract.
6. Store the extracts at −20 °C for future antibacterial and antifungal activity tests.

3.8.2 Preparation of Test Pathogens

1. As test organisms, use pathogenic bacteria and fungi, such as *Xanthomonas campestris*, *Fusarium oxysporum*, and *Botrytis cinerea*.
2. Culture bacterial pathogens in nutrient broth and fungal pathogens on potato dextrose agar.

3. Incubate the cultures at 28 °C for 24 hours to obtain active pathogens for testing.

3.8.3 Agar Well Diffusion Method

1. Evaluate actinobacterial extracts' antibacterial and antifungal activity using the agar well diffusion method.
2. Pour sterile Mueller–Hinton agar into Petri dishes for bacterial tests and potato dextrose agar for fungal tests.
3. Once solidified, inoculate the plates with the test pathogens using a sterile cotton swab to ensure even distribution.
4. Using a sterile cork borer, create wells (6 mm diameter) in the agar and add 50 μL of each actinobacterial extract to the wells.
5. Inoculate only the solvent (ethyl acetate or methanol) in the control for comparison.
6. Incubate the plates at 28 °C for 24–48 hours for bacteria and 3–7 days for fungi.

3.8.4 Assessment of Antimicrobial Activity

1. After incubation, examine the plates for clear inhibition zones around the wells, indicating antimicrobial activity.
2. Measure the diameter of inhibition zones in millimeters using a ruler or caliper.
3. Record the results and assess the effectiveness of each extract against the test pathogens.
4. Repeat the experiment in triplicate to ensure reliable and accurate results.

3.8.5 Data Analysis

1. Statistically analyze the data from the inhibition zone measurements to evaluate the significance of the antimicrobial activity of the actinobacterial extracts.
2. Select extracts showing the highest inhibition levels for further characterization and potential development as biocontrol agents against plant diseases.

4 Future Perspectives

Actinobacteria, a large and diverse group, are renowned for their ability to produce a wide range of metabolites that significantly impact agriculture and the pharmaceutical industry. Ongoing research is anticipated to reveal further beneficial traits that could improve agricultural productivity and sustainability while introducing new challenges in taxonomic reclassification and discovering innovative plant-beneficial traits with biotechnological applications. Conventional and molecular identification methods for actinobacteria have been explored, paving the way for the genetic manipulation of novel antibiotic gene clusters in heterologous hosts. Additionally, these findings may facilitate the construction of new bioactive molecules through the genetic engineering of biosynthetic pathways. These methods highlight the versatility of actinobacteria isolation techniques across various environments, facilitating their study and potential biotechnological and agricultural applications.

References

1. Augustine S, Bhavsar S, Kapadnis B (2005) Production of a growth-dependent metabolite active against dermatophytes by Streptomyces rochei AK 39. Indian J Med Res 121:164–170
2. Kurnianto MA, Kusumaningrum HD, Lioe HN (2020) Characterization of *Streptomyces* isolates associated with estuarine fish *Chanos chanos* and profiling of their antibacterial metabolites-crude-extract. Int J Microbiol 2020:1–12. https://doi.org/10.1155/2020/8851947
3. Mulya E, Waturangi DE (2021) Screening and quantification of anti-quorum sensing and antibiofilm activity of Actinomycetes isolates against food spoilage biofilm-forming bacteria. BMC Microbiol 21:1. https://doi.org/10.1186/s12866-020-02060-7
4. Wei Y, Zhao Y, Zhou D, Qi D, Li K, Tang W, Chen Y, Jing T, Zang X, Xie J, Wang W (2020) A newly isolated streptomyces sp. YYS-7 with a broad-spectrum antifungal activity improves the banana plant resistance to fusarium oxysporum f. sp. cubense tropical race 4. Front Microbiol 11:1712. https://doi.org/10.3389/fmicb.2020.01712
5. Nadar Rajivgandhi G, Ramachandran G, Li J-L, Yin L, Manoharan N, Rajesh Kannan M, Velanganni AAJ, Alharbi NS, Kadaikunnan S, Khaled JM, Li W-J (2020) Molecular identification and structural detection of the anticancer compound from marine Streptomyces akiyoshiensis GRG 6 (KY457710) against MCF-7 breast cancer cells. J King Saud Univ Sci 32:3463–3469. https://doi.org/10.1016/j.jksus.2020.10.008
6. Sousa JADJ, Olivares FL (2016) Plant growth promotion by streptomycetes: ecophysiology, mechanisms and applications. Chem Biol Technol Agric 3:24. https://doi.org/10.1186/s40538-016-0073-5
7. Rosenblueth M, Martínez-Romero E (2006) Bacterial endophytes and their interactions with hosts. MPMI 19:827–837. https://doi.org/10.1094/MPMI-19-0827
8. Goudjal Y, Zamoum M, Meklat A, Sabaou N, Mathieu F, Zitouni A (2016) Plant-growth-promoting potential of endosymbiotic actinobacteria isolated from sand truffles (Terfezia Leonis Tul.) of the Algerian Sahara. Ann Microbiol 66:91–100. https://doi.org/10.1007/s13213-015-1085-2
9. Pandey RN, Bapatla KG, Lakshmaiah M (2021) *In vitro* evaluation of actinobacteria for biocontrol of dry root rot and Fusarium wilt diseases in chickpea (Cicer arietinum L.). J Food Legum 34(4):245–247
10. Ilic SB, Konstantinovic SS, Todorovic ZB, Lazic ML, Veljkovic VB, Jokovic N, Radovanovic BC (2007) Characterization and antimicrobial activity of the bioactive metabolites in streptomycete isolates. Microbiology 76:421–428. https://doi.org/10.1134/S0026261707040066
11. Lahdenpera ML, Simon E, Uoti J (1991) Mycostop – a novel bio fungicide based on Streptomyces bacteria. In: Beemster ABR, Bollen GJ, Gerlagh M, Ruissen MA, Schippers B, Tempel A (eds) Biotic interactions and soil-borne disease. Elsevier, Amsterdam, pp 258–263
12. Mitra D, Mondal R, Khoshru B, Senapati A, Radha TK, Mahakur B, Uniyal N, Myo EM, Boutaj H, Sierra BEG, Panneerselvam P, Ganeshamurthy AN, Elković SA, Vasić T, Rani A, Dutta S, Mohapatra PKD (2022) Actinobacteria-enhanced plant growth, nutrient acquisition, and crop protection: advances in soil, plant, and microbial multifactorial interactions. Pedosphere 32:149–170. https://doi.org/10.1016/S1002-0160(21)60042-5
13. Silva GDC, Kitano IT, Ribeiro IADF, Lacava PT (2022) The potential use of actinomycetes as microbial inoculants and biopesticides in agriculture. Front Soil Sci 2:833181. https://doi.org/10.3389/fsoil.2022.833181
14. Balagurunathan R, Radhakrishnan M, Shanmugasundaram T, Gopikrishnan V, Jerrine J (2020) Evaluation of actinobacteria for aquaculture applications. In: Protocols in actinobacterial research. Springer, New York, pp 181–188. https://doi.org/10.1007/978-1-0716-0728-2_11
15. Amin DH, Abdallah NA, Abolmaaty A, Tolba S, Wellington EMH (2020)

Microbiological and molecular insights on rare actinobacteria harbouring bioactive perspective. Bull Natl Res Cent 44:5. https://doi.org/10.1186/s42269-019-0266-8
16. Elbendary AA, Hussain AM, El-Hariri MD, Seida AA, Moussa IM, Mubarak AS, Kabli SA, Hemeg HA, El Jakee JK (2018) Isolation of antimicrobial producing actinobacteria from soil samples. Saudi J Biol Sci 25:44–46. https://doi.org/10.1016/j.sjbs.2017.05.003
17. Georgousaki K, Tsafantakis N, González I, Martin J, Mackenzie TA, Gumeni S, Trougakos IP, Reyes F, Genilloud O, Fokialakis N (2025) Isolation, spectral assignments and absolute configuration of a linear hexapeptide from the culture broth of the plant-associated actinomycete actinomycetospora sp. CA-287887. Separations 12:57. https://doi.org/10.3390/separations12030057
18. Coombs JT, Franco CMM (2003) Isolation and identification of actinobacteria from surface-sterilized wheat roots. Appl Environ Microbiol 69:5603–5608. https://doi.org/10.1128/AEM.69.9.5603-5608.2003
19. Ningsih F, Sari DCAF, Rachmania MK, Yabe S, Yokota A, Oetari A, Sjamsuridzal W (2020) Isolation and 16S rRNA gene sequences analysis of thermophilic actinobacteria isolated from soil in Cisolok geothermal area, West Java, Indonesia. IOP Conf Ser: Earth Environ Sci 457:012015. https://doi.org/10.1088/1755-1315/457/1/012015
20. Chaiya L, Kumla J, Suwannarach N, Kiatsiriroat T, Lumyong S (2021) Isolation, characterization, and efficacy of actinobacteria associated with arbuscular mycorrhizal spores in promoting plant growth of chili (Capsicum flutescens L.). Microorganisms 9:1274. https://doi.org/10.3390/microorganisms9061274
21. Ali AR, Bahrami Y, Kakaei E, Mohammadzadeh S, Bouk S, Jalilian N (2022) Isolation and identification of endophytic actinobacteria from Citrullus colocynthis (L.) Schrad and their antibacterial properties. Microb Cell Factories 21:206. https://doi.org/10.1186/s12934-022-01936-9
22. Carvalho MF, Ma Y, Oliveira RS, Freitas H (2017) Endophytic actinobacteria for sustainable agricultural applications. In: Maheshwari DK, Annapurna K (eds) Endophytes: crop productivity and protection. Springer, Cham, pp 163–189. https://doi.org/10.1007/978-3-319-66544-3_8
23. Cafaro MJ, Currie CR (2005) Phylogenetic analysis of mutualistic filamentous bacteria associated with fungus-growing ants. Can J Microbiol 51:441–446. https://doi.org/10.1139/w05-023
24. Patten CL, Glick BR (2002) Role of *Pseudomonas putida* indoleacetic acid in development of the host plant root system. Appl Environ Microbiol 68:3795–3801. https://doi.org/10.1128/AEM.68.8.3795-3801.2002
25. Schwyn B, Neilands JB (1987) Universal chemical assay for detecting and determining siderophores. Anal Biochem 160:47–56. https://doi.org/10.1016/0003-2697(87)90612-9
26. Mehta S, Nautiyal CS (2001) An efficient method for qualitative screening of phosphate-solubilizing bacteria. Curr Microbiol 43:51–56. https://doi.org/10.1007/s002840010259
27. Su Y, Wang Z, Liu F, Li Z, Peng Q, Guo J, Xu L, Que Y (2016) Isolation and characterization of ScGluD2, a new sugarcane beta-1,3-glucanase D family gene induced by sporisorium scitamineum, ABA, H2O2, NaCl, and CdCl2 stresses. Front Plant Sci 7. https://doi.org/10.3389/fpls.2016.01348
28. Cappuccino JG, Sherman N (2005) Microbiology: a laboratory manual. Pearson/Benjamin Cummings, San Francisco, p 507
29. Chukwuneme CF (2018) Actinomycetes impacts on drought stress in maize (*Zea Mays* L.) (Doctoral dissertation, North-West University (South Africa))
30. Rodríguez H, Fraga R (1999) Phosphate solubilizing bacteria and their role in plant growth promotion. Biotechnol Adv 17:319–339. https://doi.org/10.1016/S0734-9750(99)00014-2
31. Djebaili R, Pellegrini M, Smati M, Del Gallo M, Kitouni M (2020) Actinomycete strains isolated from saline soils: plant-growth-promoting traits and inoculation effects on solanum lycopersicum. Sustainability 12:4617. https://doi.org/10.3390/su12114617
32. Zaidi A, Khan MS, Ahmad M, Oves M, Wani PA (2009) Recent advances in plant growth promotion by phosphate-solubilizing microbes. In: Khan MS, Zaidi A, Musarrat J (eds) Microbial strategies for crop improvement. Springer, Berlin/Heidelberg, pp 23–50. https://doi.org/10.1007/978-3-642-01979-1_2

33. Singh T, Wahla V, Dubey RC, Maheshwari DK (2022) Destructive mycolytic suppression of Fusarium oxysporum causing wilt in chickpea by fungicide tolerant actinobacteria. Environ Sustain 5:243–253. https://doi.org/10.1007/s42398-022-00229-6

34. Cao L, Qiu Z, You J, Tan H, Zhou S (2004) Isolation and characterization of endophytic Streptomyces strains from surface-sterilized tomato (Lycopersicon esculentum) roots. Lett Appl Microbiol 39:425–430. https://doi.org/10.1111/j.1472-765X.2004.01606.x

Chapter 8

Isolation of Endosymbionts from Green Algae

P. V. Tharani and K. V. Bhaskara Rao

Abstract

This chapter discusses the isolation of endosymbiotic microorganisms such as bacteria, actinobacteria, and fungi from green algae. Several methods for collecting and sterilizing algal samples are discussed, ensuring that the isolated microorganisms are truly endosymbiotic. Each species of endosymbionts is cultured on a different medium like nutrient agar (NA) for bacteria, actinomycete isolation agar (AIA) and International Streptomyces Project Medium (ISP-2) for actinobacteria, and potato dextrose agar (PDA) with streptomycin for fungi. Successful endosymbiont isolation requires adequate sample handling, sterilization techniques, and appropriate conditions to encourage the growth of desired microorganisms while inhibiting contaminants.

Key words Green algae, Endosymbionts, Endophytes, Culture media

1 Introduction

Green algae, a wide category of photosynthetic organisms, are categorized as green lineage (Chloroplastida) within the class Archaeplastida [1]. Streptophyta, the closest relatives of terrestrial plants, share several characteristics with them, including the existence of phragmoplasts and oogamous reproduction. Prasinodermophyta, a fundamental group of green algae, is distinguished by its small, unicellular members. Chlorophyta, the largest and most diversified group, includes a vast variety of morphological forms, ranging from unicellular flagellates to multicellular seaweed. Green algae are classified based on their phylogenetic links and shared traits, which represent their evolution and diversity [2]. Endophytes are endosymbionts, and they are mostly of bacterial and fungal origin. These are the microorganisms that colonize intercellular spaces in plants in natural habitats. They might be bacteria, actinobacteria, or fungi. In recent years, their secondary metabolites have received widespread recognition for their medical and industrial applications [3]. Endosymbionts, which are microorganisms that dwell within plant tissues without causing harm, can be isolated

N. Amaresan and Sapna Chandwani (eds.), *Endosymbionts*, Springer Protocols Handbooks,
https://doi.org/10.1007/978-1-0716-4771-4_8,

using a variety of microbiological media, including nutrient agar, potato dextrose agar, and nitrogen- or carbon-based medium. Endosymbionts are frequently identified and enumerated by isolating them from surface-sterilized host plant tissue. Endophytes are endosymbiotic microorganisms that colonize within a plant or microbial species and are influenced by a variety of parameters such as the plant's genotype, growth stage, physiological status, tissue type, ambient circumstances, sampling season, surface sterility, selective media, and culture conditions. These bacteria produce a diverse array of bioactive metabolites, such as phenolic acids, alkaloids, quinones, steroids, saponins, tannins, and terpenoids. These bacteria produce a diverse array of bioactive metabolites, such as phenolic acids, steroids, saponins, alkaloids, quinones, tannins, and terpenoids. These metabolites have demonstrated potential for a variety of medical uses, including anticancer, antimalarial, antitubercular, antiviral, anti-diabetic, anti-inflammatory, antiarthritic, and immunosuppressive characteristics. While the primary role of these bioactive chemicals is to improve host plant resilience to abiotic and biotic stressors, endophytes are nevertheless a unique source of potential drug discovery [3, 4].

2 Materials

All solutions should be prepared with sterile distilled water and analytical-grade reagents.

2.1 Sample Collection

1. Plastic boxes or sterile polythene zip-lock containers.
2. Double-distilled water.

2.2 Endosymbiotic Bacterial Isolation

1. 70% ethanol: for a 100 mL solution, 70 mL ethanol, and 30 mL water.
2. Sterile autoclaved distilled water.
3. Nutrient agar (NA).

2.3 Endosymbiotic Actinobacterial Isolation

1. 70% ethanol.
2. Sodium hypochlorite: solution of 1 part sodium hypochlorite to 10 parts water (*see* **Note 1**).
3. Sterile double-distilled water.
4. Actinomycete isolation agar.
5. International Streptomyces Project Medium (ISP-2) agar.

2.4 Endosymbiotic Fungal Isolation

1. Sterile double-distilled water.
2. 70% ethanol.

3. 1.0% sodium hypochlorite.
4. Potato dextrose agar.

3 Methods

3.1 Sample Collection

3.1.1 Method 1

1. Transport the algae to the laboratory in sealed plastic boxes containing the water where they were collected. This helps to preserve the algae's natural environment throughout shipping.
2. To keep the algae from decaying, keep the containers chilled at around 25 degrees celsius.
3. Transfer the algal cultures to the laboratory. Keep the algae in their original containers, and air should be bubbled through the water to maintain their health and viability until the isolation process begins [5] (*see* **Note 1**).

3.1.2 Method 2

1. To avoid contamination, keep the algal samples in a sterile zip-lock bag. Bring the material to the laboratory in an ice box.
2. Clean the algal sample thoroughly with tap water to eliminate free-floating organisms such as epiphytes.
3. Surface sterilize the samples twice with double-distilled water for 5 min, then keep them at 4 °C for future use [6].

3.1.3 Method 3

1. Clean the seaweed samples with seawater to remove salts and debris before transporting them to the laboratory and store the samples in sterile polythene zip-lock containers containing seawater.
2. To avoid contamination, transport the samples to the laboratory aseptically.
3. Isolate the endosymbionts from marine macroalgae after the taxonomic identification of algae [7].

3.2 Endosymbiotic Bacterial Isolation

1. Wash the collected algal samples properly with fresh water to eliminate any residual debris.
2. Soak the small pieces of algae in 70% ethanol for 3 s and then rinse twice in distilled water for 1 min each to ensure surface sterilization (*see* **Note 2**).
3. Ensure the proper sterilization of the algal segments by inoculating water samples from the final rinse onto appropriate agar media. The absence of bacterial or fungal development will ensure the sterility of the algal surfaces.
4. Cut the sterilized algal fragments into roughly 0.5 cm squares and place them in 90-mm Petri plates with nutrient agar (NA) for bacteria.

5. Incubate the multiple plates of each media, each with ten algal pieces at 25 °C for 8 days to allow endophytic bacteria to develop. Then, pick up the bacterial colonies and re-inoculate them on suitable growth media. To maintain culture purity, streak each single colony to obtain a pure culture [6].

3.3 Endosymbiotic Actinomycete Isolation

1. Isolate endophytic actinomycetes from green algae using actinomycete isolation agar (AIA).
2. Slice algal samples aseptically into 10-mm segments and sterilize them with 70% ethanol and sodium hypochlorite.
3. To remove any remaining surface contamination, wash the samples thoroughly with sterile double-distilled water.
4. Add the sterilized algal segments to actinomycete isolation agar (AIA) plates and culture them at 28 °C for 5–10 days (*see* **Note 3**).
5. Choose the colonies with powdery white, greyish white, or pale yellow-white pigmentation and preserve them on starch casein agar plates at 4 °C for future examination.
6. To confirm the endophytic nature of the recovered actinomycetes, directly inoculate the unsterilized algal samples and the final wash water onto International Streptomyces Project Medium (ISP-2) agar plates.
7. The absence of microbial growth on ISP-2 plates inoculated with wash water is to ensure that there is no epiphyte contamination. Furthermore, discover the surface-contaminating microorganisms by macroscopic and microscopic analysis of unsterilized algal samples [6].

3.4 Endosymbiotic Fungal Isolation

1. Clean the seaweed sections (about 2 × 2 cm) with sterile double-distilled water, then submerge them in 70% ethanol for 60 s and then in 1.0% sodium hypochlorite for 30 s.
2. After a final 30-s wash in 70% ethanol, rinse the samples with sterile double-distilled water and blot dry the sample.
3. After surface sterilization, cut the algal samples into pieces and inoculate them into potato dextrose agar (PDA) medium with 250 mg/L streptomycin to inhibit bacterial growth (*see* **Note 4**).
4. To guarantee that only endophytic fungi are isolated, inoculate or plate the rinse water on PDA to observe epiphyte contamination. Incubate the cultivated plates at 25 ± 2 °C for 10 days. Transfer the endophytic fungi isolated from marine seaweed to new PDA plates and allow them to grow separately for future investigations [7, 8].

4 Notes

1. For algae with thin surfaces, use a diluted solution of 1 part sodium hypochlorite to 10 parts water, and for algae with thick surfaces, use a stronger solution of 1 part sodium hypochlorite to 5 parts water.
2. The surface being treated with ethanol or sodium hypochlorite may also influence the concentration and manner of application. For example, porous surfaces may necessitate a stronger solution or more applications.
3. ISP2 and AIA are two types of culture media preferred to isolate actinomycetes. They have the same composition except that ISP2 does not have antifungal or antibacterial agents, whereas AIA has antifungal and antibacterial agents, which is helpful in the isolation of actinomycetes and devoid of other bacterial and fungal growth.
4. Even though PDA is most commonly used for fungal isolation, several other media can also be opted to collect diverse fungal groups; for example, yeast malt agar favors yeast.

Acknowledgement

The authorities and employees of VIT University, Vellore, Tamil Nadu, India, deserve our deepest gratitude for supporting our initiatives.

References

1. Bowles AMC, Paps J, Bechtold U (2022) Water-related innovations in land plants evolved by different patterns of gene cooption and novelty. New Phytol 235:732–742
2. Bachy C, Wittmers F, Muschiol J, Hamilton M, Henrissat B, Worden AZ (2022) The land–sea connection: insights into the plant lineage from a green algal perspective. Annu Rev Plant Biol 73:585–616
3. Gouda S, Das G, Sen SK, Shin HS, Patra JK (2016) Endophytes: a treasure house of bioactive compounds of medicinal importance. Front Microbiol (Frontiers Media SA) 7
4. Fadiji AE, Babalola OO (2020) Elucidating mechanisms of endophytes used in plant protection and other bioactivities with multifunctional prospects. Front Bioeng Biotechnol 8
5. Deutsch Y, Gur L, Berman Frank I, Ezra D. Endophytes from algae, a potential source for new biologically active metabolites for disease management in aquaculture. Front Mar Sci 2021;8
6. Ramachandran G, Rajivgandhi G, Maruthupandy M, Manoharan N (2019) Extraction and partial purification of secondary metabolites from endophytic actinomycetes of marine green algae Caulerpa racemosa against multi drug resistant uropathogens. Biocatal Agric Biotechnol 17:750–757
7. Venkatachalam J, Jeyadoss VS, Bose KSC, Subramanian R (2024) Marine seaweed endophytic fungi-derived active metabolites promote reactive oxygen species-induced cell cycle arrest and apoptosis in human breast cancer cells. Mol Biol Rep 51:611
8. Blessie EJ, Wruck W, Abbey BA, Ncube A, Graffmann N, Amarh V et al (2020) Transcriptomic analysis of marine endophytic fungi extract identifies highly enriched anti-fungal fractions targeting cancer pathways in HepG2 cell lines. BMC Genomics 21

Chapter 9

Isolation and Identification of Culturable Endosymbiotic Bacteria from Insects

Satish V. Patil, Bhavana V. Mohite, Jitendra D. Salunkhe, and Sunil H. Koli

Abstract

Insects are the most dominant and successful group in the animal kingdom. This may be due to their highly active defense system, which protects them from various biotic and abiotic factors. Besides these, most insects maintain mutualistic relationships with certain microbes that reside in their bodies, known as endosymbionts. These microbes play vital roles ranging from nutritional supplementation to defense against both abiotic and biotic factors. Endosymbionts are primarily categorized into two types: primary and secondary. Primary endosymbionts are vertically transmitted from mother to offspring, and their roles are considered critical for the nutrition, survival, and development of the host. Most of the obligatory symbionts cannot be cultured on artificial media. Even culturing or isolating the culturable endosymbionts presents a significant challenge. The technique proposed in this protocol for isolating insect endosymbionts involves using specific media and a sandwich method to provide the essential environment required for their growth.

Key words Endosymbionts, Symbiosis, Nutrition, Reproduction, Resistance

1 Introduction

The term symbiosis was coined by Anton de Bary in 1879, referring to the presence of one organism within another without causing harm [1]. Such associations have been studied across the kingdoms of life, with many organisms identified as a major force driving the evolution of life [2–4]. Mereschkowsky (1910) proposed that the evolution of the eukaryotic cell originated from the prokaryotic cells [5]. The roles of endosymbiotic microbes have been recognized in various insects, ranging from reproductive modulation and essential amino acid production to contributing to eukaryotic evolution and pesticide resistance [3, 6–8]. Moran et al. (2008) reported that *Phytosuccivorous hemipteran* insects essentially rely on intracellular symbiotic microbes to fulfill their amino acid requirements [9]. Recently, Jing et al. (2020) demonstrated that *Cryptorhynchus lapathi* harbors dominant gut microbes essential for

N. Amaresan and Sapna Chandwani (eds.), *Endosymbionts*, Springer Protocols Handbooks,
https://doi.org/10.1007/978-1-0716-4771-4_9,

food digestion and nutrition [10]. The presence of various bacterial endosymbionts has also been reported in common Indian insects. Sharma and Jaiswal (2011) identified *Micrococcus varians*, *M. conglomeratus*, *Clostridium* sp., and *Bacillus subtilis* as endosymbionts in various Indian insect pests [11].

The endosymbionts are broadly classified into two types: primary and secondary. Primary endosymbionts are vertically transmitted from mother to offspring and play a critical role in the nutrition, survival, and development of the host. Obligate endosymbiotic bacteria such as *Buchnera aphidicola* and *Wigglesworthia glossinidia* are essential for supplying vitamins to their respective hosts, the pea aphid (*Acyrthosiphon pisum*) and tsetse fly. In addition to primary symbionts, many insects also harbor secondary endosymbionts, which can be transmitted both vertically and horizontally and typically found in the hemocoel. For example, Matthew et al. (2005) for the first time detected and cultured the bacteria *Sodalis glossinidius* from the hemocoel of the tsetse fly using an artificial medium [12]. Secondary symbionts are generally cultivable in artificial media under laboratory conditions. The technique proposed in this protocol for isolating insect endosymbionts involves the use of specific media and a sandwich method, which provides an optimal environment for bacterial growth [8].

2 Materials

1. Insects sample/IV instar larvae.
2. For bacterial endosymbionts: minimal nutrient medium (peptone: 1.25 g/L, yeast extract: 1.25 g/L, dextrose: 3.0 g/L, agar: 30.0 g/L).
3. For fungal endosymbiont isolation: potato dextrose agar medium (PDA) fortified with tetracycline 120 μg/L and streptomycin 150 μg/L.
4. For yeast endosymbionts: YPD agar (yeast extract: 10 g/L, peptone: 20 g/L, and dextrose: 20 g/L).
5. For protozoan endosymbionts: protozoan minimal medium (peptone: 1.5 g/L, NaCl: 0.5 g/L, and yeast extract: 0.5 g/L) (*see* **Note 1**).
6. Media for nematode endosymbionts: for insect nematode endosymbionts, sterile media containing agar–agar: 20 g/L with yeast extract: 20 g/L.
7. Base agar: agar–agar 15–20 g/L in distilled water.
8. Sterile saline tubes each containing 5 mL sterile saline.
9. Sterile Petri plates.
10. *E. coli* culture.

11. Autoclave.
12. Heating chamber.
13. Microscope.
14. Staining reagents.
15. Sterile microsyringes.

3 Methods

3.1 Collection of Insect and Larvae Samples

1. Collect insects or insect larvae as required, and place them in sterile plastic Petri plates.
2. Seal the plates or transfer them into sterile plastic bags.
3. Label each sample appropriately with relevant information such as location, insect species, and date of collection (*see* **Note 2**).
4. Store the samples at 4 °C until further use in experimental procedures (*see* **Note 3**).

3.2 Preparation of Sample

1. To isolate the endosymbionts from insects or insect larvae, first freeze the samples at 4 °C for 10–15 min.
2. Next, perform surface sterilization by sequentially rinsing the frozen insects or larvae for 1 min each in the solutions of sodium hypochlorite (1%), sterile phosphate-buffered saline (PBS), and ethanol (70%).
3. After these treatments, rinse the insects repeatedly in sterile phosphate buffer for 10 min.
4. To confirm the effectiveness of surface sterilization, take 100 μL of the final phosphate buffer rinse and inoculate it into a minimal nutrient agar butt.
5. Pour this onto base agar and treat it as a control sample.
6. Next, surface sterilize the specimen again by immersing them in 70% ethanol, followed by washing with sterile distilled water.
7. Repeat this sterilization and washing procedure twice.
8. All operations should be conducted under a laminar airflow hood to maintain sterility.
9. Transfer the sample to sterile Petri plates and incubate at 4 °C for 10 min.
10. Apply the same procedure to insect larvae, especially those at specific instar stages.
11. Carefully dissect the larvae to expose the digestive tract or extract hemolymph directly from the hemocoel using a micro-syringe for endosymbiont isolation.

12. This method is most suitable for large insects or larval stages such as *Megasoma elephas*, *Megasoma actaeon*, and *Goliathus goliatus* (*see* **Note 4**).
13. Following dissection, isolate the gastrointestinal system and add sterile saline using a sterile glass rod.
14. Gently squeeze the tissue and homogenize the sample (*see* **Note 5**).
15. The extracted hemolymph or homogenate may then be used for the isolation of organisms using the sandwich methods (*see* **Notes 5** and **6**).
16. Observe the cultivated endosymbiotic protozoal cells using a wet mount preparation on a cavity slide covered with a coverslip under 40x.
17. Record morphological features of the observed structures and identify the endosymbionts at both protozoal cysts and trophozoite stages using simple staining methods, such as methylene blue, cotton blue, and Giemsa staining (*see* **Note 7**).

3.3 Culture Medium Preparation for Protozoa

1. Protozoa can be cultivated on non-nutrient agar or agar media containing minimal concentrations of nutrients.
2. Sterilize the medium by autoclaving at 15 lbs. pressure (121 °C) for 20 min.
3. Allow the medium to cool to approximately 45 °C, then pour into sterile Petri plates, and allow them to solidify for 1 h at room temperature or 2 h at 30 °C.
4. After solidification, aseptically spread the test sample and incubate for 2 h, then overlay with 100 μL of a 24 h old culture of attenuated *E. coli* (prepared by heating at 60 °C for 10 min).
5. Incubate the plates at room temperature for 24 h (*see* **Note 8**).

3.4 Isolation and Cultivation of Endosymbionts of Insects

3.4.1 Part I: Isolation of Endosymbionts on Solid Media

1. Preparation of base agar

   Aseptically pour sterile base agar into sterile Petri plates and allow it to solidify at room temperature.
2. Sample dilution

   Dilute the prepared insect or larvae lysate using sterile saline through routine serial dilution techniques up to 10^{-9} dilution and mix each dilution thoroughly.
3. Inoculation of diluted sample

   Aseptically transfer 0.5 mL from the 10^{-5} or 10^{-6} dilution into sterile agar butts containing the appropriate culture medium, depending on the target organism.

For protozoa: Use protozoan-specific medium.

For bacteria: Use the minimal nutrient medium.

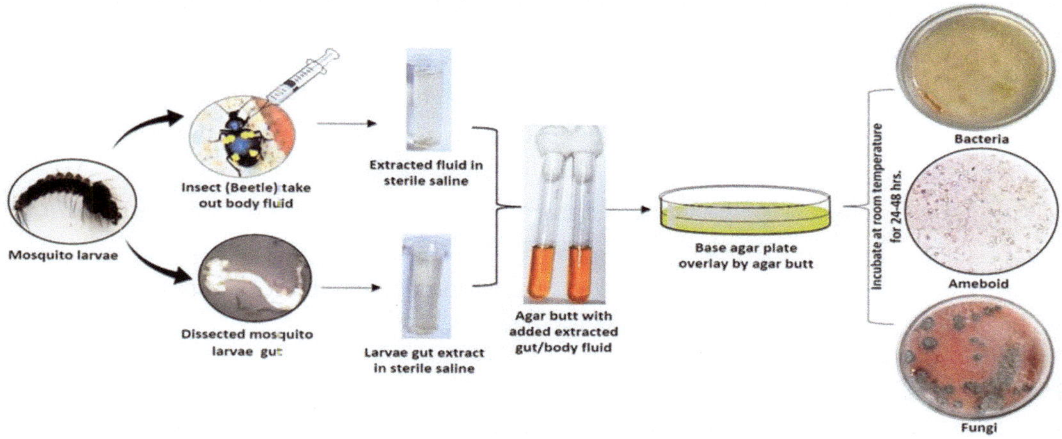

Fig. 1 General protocol for isolations of endosymbionts of insects

For nematode: Use nematode minimal agar.

For yeast and fungi: Use a YPD medium.

Mix the sample by gentle vertexing, and overlay it onto the surface of the base agar plate (*see* **Note 6** and Fig. 1). Gently rotate the plate clockwise and anticlockwise to ensure uniform spreading.

4. Protozoan culture step
   For protozoan isolation, after a 2 h incubation, aseptically overlay 100 μL of attenuated *E. coli* culture.
5. Incubation condition
   For bacteria: Incubate at 37 °C for 24–48 h.
   For yeast and fungi: Incubate at 30 °C for 48–72 h.
   For protozoa: Incubate at 30 °C for 24 h.

3.4.2 Part II: Observation of Growth

1. The culturable endosymbionts isolated from insects are generally microaerophilic and tend to grow better under submerged conditions.
2. Following incubation, observe the culture plates under a light microscope at 40x to identify colonies of bacteria, fungi, and yeast.
3. Generally, trophozoites of protozoa or free-living nematodes are typically visible as actively motile forms within the agar medium.

3.5 Primary Identification (Staining)

Colonies of culturable endosymbionts often develop in the subsurface layers of the agar. Using a sterile straight wire loop, carefully excise a portion of the agar containing visible or suspected colonies and suspend it into sterile saline for further processing.

3.5.1 Preliminary Bacterial Observation (Negative Staining)

1. Take a loopful of the prepared suspension on one end of a clean glass slide.
2. Add a drop of Indian ink to the suspension and gently mix.
3. Using another slide held at an angle, spread the drop to form a thin smear and allow it to air-dry.
4. Examine the smear under 40x to assess the presence and morphology of bacterial cells.
5. Note the size, shape, and arrangement of unstained bacterial cells against the dark background.

3.5.2 Differential Staining (Gram Staining)

1. Once bacterial morphology is observed, proceed with Gram staining for preliminary classification.
2. This differential stain helps to distinguish between Gram-positive and Gram-negative bacteria.

3.5.3 Yeast and Fungi Staining

1. Prepare a suspension from colonies grown on YPD medium.
2. Take a loopful of the suspension and mix it with a drop of cotton blue on a clean glass slide.
3. Incubate at room temperature for 5 min.
4. Place a coverslip over the preparation, taking care to avoid air bubbles.
5. Observe the slide under a light microscope using the 40x objective.
6. Record the morphology of hyphae and spores.

3.5.4 Nematode Staining

1. Excise a small agar block from nematode minimal agar containing visible nematodes and suspend it in sterile saline.
2. Using a Pasteur pipette under a dissecting microscope, isolate individual nematodes and transfer them into a sterile Petri plate.
3. Add a drop of saline containing isolated nematodes along with a few drops of formalin using a Pasteur pipette.
4. Prepared a wet mount on a glass slide and observe under 40x objective (*see* **Note 3**).
5. For staining, treat the nematodes with trypsin to facilitate dye uptake (*see* **Note 8**).
6. Following enzymatic treatment, stain the nematodes with a few drops of neutral red (0.005%) or eosin Y dye.
7. Incubate the stained preparation for 5 min and observe nematodes under a 10x objective to assess morphology.

3.6 Identification and Observations of Protozoa

1. At every 24 h, take a loopful of the culture and dilute it with a drop of sterile water.
2. Place the diluted drop carefully at the center of a cavity slide. Apply wax at the four corners of the slide and invert it to create a hanging drop preparation. Observe under the microscope using the 40x objective.
3. Record the shape and structure of the cultivated protozoa.
4. For further observation, use staining techniques such as methylene blue, cotton blue, iodine, and Giemsa staining (*see* **Note 7**).

Staining of Protozoa

1. Take a loopful of the protozoan culture and spread it onto a clean glass slide. Gently heat the slide over a flame to fix the smear.
2. Flood the fixed smear of the protozoan cyst with freshly filtered carbol fuchsin solution.
3. Rinse the smear thoroughly under tap water and then flood it with malachite green as a counter stain for 10 s.
4. Rinse the smear again with tap water, air-dry, and observe under the microscope.

3.7 Part III: Axenic Culture Preparation of Endosymbionts

3.7.1 Preparation of Pure Culture of Bacterial Endosymbionts

1. Cut the agar block containing a single bacterial colony from the plate and suspend it in 5 mL sterile saline.
2. Perform serial dilutions of the suspension and inoculate 0.1 mL from the 10^{-6} dilution into sterile minimal nutrient agar butts. Mix well and pour onto base agar plates. Incubate the plates at 37 °C for 24 h.
3. Confirm colony growth by Gram staining and compare with initial morphological observations.
4. Further identification can be performed using molecular techniques such as 16S rRNA gene sequencing.

3.8 Preparation of Pure Culture of Yeast and Fungal Endosymbionts

1. Cut the agar block containing a single colony from a YPD plate and suspend it in 5 mL sterile saline.
2. Serially dilute the suspension and inoculate 0.1 mL of the 10^{-6} dilution into sterile YPD agar butts. Mix well and pour onto the base agar plate. Incubate the plates at 28 °C for 48 h.
3. Confirm the identity of the yeast or fungi colonies using lactophenol cotton blue staining and compare with initial morphological features.
4. Further identification should be done using molecular methods such as 18S rRNA or ITS region sequencing.

3.9 Preparation of Pure Culture of Protozoan Endosymbionts

1. Cut the agar block from the protozoan culture plate and inoculate it into peptone–yeast extract–dextrose (PYD) broth and incubate for 8 days at 30 °C.
2. Harvest the cysts by centrifugation at 5000 rpm. Resuspend the pellet in 5 mL of sterile saline containing 0.5% sodium dodecyl sulfate (SDS) (*see* **Note 9**). Centrifuge again and collect the pellet.
3. Incubate the pellet overnight in 3% HCl to eliminate any remaining bacterial contamination.
4. Wash the cysts two to three times with physiological saline (centrifuge at 7000 rpm for 10 min) to remove the remaining HCl. Then, transfer the cleaned cysts to a sterile PYD medium.
5. Maintain axenic cultures by subculturing into fresh sterile PYD every 2 weeks.

3.10 Preparation of Pure Culture of Nematode Endosymbionts

1. Cut the agar block from nematode minimal agar and suspend it in sterile saline.
2. Using Pasteur pipettes, isolate single egg–bearing nematodes based on morphological features and transfer them to sterile Petri plates containing saline.
3. Inoculate individual nematodes into fresh nematode agar plates and incubate for 1 week at 25 °C.
4. Identify nematode species using morphological observations and molecular techniques.

4 Notes

1. Protozoa require bacteria as a nutrient source. To support growth, overlay 100 μL of a 24-h-old attenuated *E. coli* culture (heated at 60 °C for 10 min) onto the plate after 2 h spreading the insect homogenate. Incubate the plates for 1 h at room temperature.
2. Endosymbionts such as protozoa, nematodes, and bacteria may include pathogenic and nonpathogenic species. Therefore, handle all samples using appropriate biosafety precautions. Prefer collecting samples from minimally disturbed natural environments.
3. Dissecting live insects or larvae is challenging. Freezing them at 4 °C for 15–30 min renders them sluggish but alive, allowing easier dissection without damaging internal organs.
4. Larger insects such as *Megasoma elephas*, *Megasoma actaeon*, *Goliathus goliatus*, *Deinacrida heteracantha* (giant weta), *Helicoverpa armigera*, butterflies, and dragonflies are ideal sources for hemocoel fluid collecting using sterile syringes for endosymbiont isolation.

5. The extract of the digestive system in sterile saline or collected fluid can also be directly used for molecular detection and identification of microbes via sequencing methods.
6. The sandwich method is essential for cultivating endosymbionts. Pour the inoculated culture medium butt aseptically onto a solidified sterile base agar layer (e.g., 3% agar–agar in 100 mL distilled water) to simulate a microaerophilic environment.
7. Eosin stain for wet mount stain: 1 g eosin in 100 mL distilled water with 0.5 mL acetic acid.

   Cotton blue/methylene blue stain: Dissolve 1 g stain in 100 mL physiological saline.

   Carbol fuchsin and iodine: may also be used for staining of protozoan cysts.
8. For nematode staining, wash formalin-treated nematodes two to three times with saline. Add 100 μL of trypsin solution (100 μg/mL) and incubate for 5 min. Apply appropriate stains after trypsin treatment.
9. SDS (0.5% final concentration) is used to confirm the transformation of trophozoites into cysts. Trophozoites are SDS-sensitive, and any remaining should be lysed immediately upon the addition of SDS, while cysts remain intact [13].

Acknowledgements

Prof. Satish V Patil acknowledges Prof. Laxmikant V Shinde, JES College, Jalna, Maharashtra for his help to increase interest in entomology.

References

1. de Bary A (1879) Die Erscheinung der Symbiose:Vortrag gehalten auf der versammlung deutscher naturforscher und aerzte zu cassel. Strassburg, Verlag Von Karl J. Trubner
2. Douglas AE (2014) Symbiosis as a general principle in eukaryotic evolution. Cold Spring Harb Perspect Biol 6:a016113
3. López-García P, Eme L, Moreira D (2017) Symbiosis in eukaryotic evolution. J Theor Biol 434:20–33
4. Heddi A, Zaidman-Rémy A (2018) Endosymbiosis as a source of immune innovation. C R Biologies 341:290–296
5. Mereschkowsky C (1910) Theorie der zwei Plasmaarten als Grundlage der Symbiogenesis, einer neuen Lehre von der Entstehung der Organismen. Biol Centralbl 30:278–288
6. Verma S, Ramani R, Sachan A, Chandra R (2023) The role of Wolbachia and the environment on sex determination of the Indian lac insect, Kerria lacca (Coccoidea: Tachardiidae). J Asia Pac Entomol 26:102019
7. Kandasamy T, Ekbal S, Kumari K et al (2022) Unraveling bacterial diversity of the Indian lac insect Kerria lacca (Kerr) using next generation sequencing. Int J Trop Insect Sci 42:2365–2372
8. Patil CD, Borase HP, Salunke BK, Patil SV (2013) Alteration in Bacillus thuringiensis toxicity by curing gut flora: novel approach for

mosquito resistance management. Parasitol Res 112:3283–3288

9. Moran NA, McCutcheon JP, Nakabachi A (2008) Genomics and evolution of heritable bacterial symbionts. Annu Rev Genet 42:165–190
10. Jing T-Z, Qi F-H, Wang Z-Y (2020) Most dominant roles of insect gut bacteria: digestion, detoxification, or essential nutrient provision? Microbiome 8:38
11. Sharma K, Jaiswal A (2011) Microbial Flora associated with lac insects. In: Indian Lac Research Institute (ed) Recent advances in lac insects, 2nd. Namkum, Ranchi, pp 22–33
12. Matthew CZ, Darby AC, Young SA et al (2005) The rapid isolation and growth dynamics of the tsetse symbiont *Sodalis glossinidius*. FEMS Microbiol Lett 248:69–74
13. Dudley R, Matin A, Alsam S et al (2005) Acanthamoeba isolates belonging to T1, T2, T3, T4 but not T7 encyst in response to increased osmolarity and cysts do not bind to human corneal epithelial cells. Acta Trop 95: 100–108

Chapter 10

Isolation and Maintenance of Endosymbionts from Mycorrhiza

Shimpy Sarkar, Arshdeep Singh, Shivam Singh, and Gagan Kumar

Abstract

Mycorrhizas are able to colonize a wide range of environments, and this ecological success is indicative of the great diversity in the fungal endophytes' genetic and physiological makeup. Fungal hyphae enter root cells in endomycorrhizas and form an internal symbiosis; in ectomycorrhizas, the hyphae stay extracellular. Microbes with genes that may affect the metabolism of bacteria, fungi, and plants include endosymbionts. The mycorrhizal fungus and nitrogen-fixing bacteria found in nature may make a great biofertilizer to increase crop yields while reducing the adverse effects of artificial fertilizers. In return for nutrients and water, mycorrhizal fungus receives over 20% of the carbon fixed by plants during photosynthesis. The more prevalent symbiosis, known as endomycorrhiza, is found in agricultural areas and is linked to over 85% of plants. Since they do not alter the shape of the roots like ectomycorrhiza do, they are more difficult to identify at first glance. Endomycorrhiza is of three different types: arbuscular, orchid, and ericloid. The separation and upkeep of mycorrhizal endosymbionts are the main topics of this chapter.

Key words Endosymbionts, Biofertilizer, Physiological, Internal symbiosis, Ecological success, Isolation, Internal symbiosis

1 Introduction

Endosymbionts, microorganisms that live within the tissues of a host organism in a symbiotic relationship, play a crucial role in the health and functionality of mycorrhizal associations. Mycorrhizae, the symbiotic relationships between fungi and plant roots, are essential for nutrient exchange, enhancing plant growth, and increasing resistance to environmental stresses. The isolation and maintenance of endosymbionts from mycorrhizal fungi are critical for understanding and harnessing these symbiotic relationships for agricultural and ecological benefits. Endosymbionts within mycorrhizal fungi, including bacteria and other microorganisms, contribute to the overall efficiency and resilience of mycorrhizal associations. They aid in nutrient cycling, protect against pathogens, and can even influence the establishment and functioning of

N. Amaresan and Sapna Chandwari (eds.), *Endosymbionts*, Springer Protocols Handbooks,
https://doi.org/10.1007/978-1-0716-4771-4_10,

mycorrhizal networks. The study of these endosymbionts offers insights into the complex interactions that underpin plant–fungal symbiosis and holds potential for the development of sustainable agricultural practices.

However, isolating and maintaining endosymbionts from mycorrhiza presents significant challenges. The intricate and often delicate nature of these associations requires precise techniques to ensure the viability and functionality of both the fungi and their endosymbionts. Various methods have been developed to isolate these microorganisms, ranging from traditional culturing techniques to advanced molecular approaches. Once isolated, maintaining these endosymbionts in vitro requires careful attention to their specific growth requirements and symbiotic needs.

This chapter provides an overview of the importance of endosymbionts in mycorrhizal associations, the challenges involved in their isolation and maintenance, and the methods employed to study these vital microorganisms. By understanding and preserving endosymbionts, researchers can unlock new possibilities for enhancing plant health and productivity through optimized mycorrhizal symbiosis.

2 Materials

2.1 Laboratory Equipment and Supplies

- *Sterilization Equipment:* autoclave, Bunsen burner, or laminar flow hood.
- *Microscopy*: stereoscopic microscope and compound microscope.
- *Growth Chambers:* for cultivating host plants and mycorrhizal fungi.
- *Incubators:* for controlled temperature and humidity conditions.
- *Centrifuges:* for separating different components of the sample.
- *pH Meters:* to measure the pH of growth media.
- *Autoclave bags, Petri dishes, Erlenmeyer flasks, test tubes, pipettes, and other standard laboratory glassware.*
- *Filters:* for sterilizing solutions and air.
- *Balances:* for weighing materials [1, 2].

2.2 Chemicals and Reagents

- *Growth Media:* for cultivating host plants and mycorrhizal fungi (e.g., Murashige and Skoog medium, vermiculite, perlite).
- *Sterilization Agents:* autoclave tape, ethanol, and bleach.
- *Buffers*: for maintaining pH and osmotic conditions (e.g., phosphate buffer).
- *Nutrients*: for fungal growth (e.g., glucose, yeast extract).

- *Antibiotics*: to prevent bacterial contamination (e.g., streptomycin, ampicillin).
- *Hormones*: for plant growth regulation (e.g., auxins, cytokinins) [1, 2].

2.3 Biological Materials

- *Host Plants*: suitable for establishing mycorrhizal associations (e.g., tomato, maize, *Medicago*).
- *Mycorrhizal Inoculum*: obtained from soil or commercial sources.
- *Endosymbiont Cultures*: if available, to initiate pure cultures.
- *Safety Equipment*: lab coats, gloves, safety goggles, and appropriate waste disposal containers [1, 2].

2.4 Specific Requirements

The exact materials needed depend on the specific research question and the type of endosymbiont we are studying. For example

- If we are working with arbuscular mycorrhizal fungi (AMF), we might require specific growth conditions and media.
- If we are studying ectomycorrhizal (ECM) fungi, different growth substrates and incubation conditions may be necessary.
- Isolation and maintenance of bacterial or fungal endosymbionts from mycorrhizal roots might involve additional techniques and materials [1, 2].

3 Methods

3.1 Isolation of Endosymbionts from Mycorrhiza

To guarantee the separation of the symbiotic organisms from the mycorrhizal fungus, there are multiple processes involved in the isolation of endosymbionts from mycorrhiza [3, 4]:

1. *Sample Collection*: Gather roots from plants that have been shown to have mycorrhizal relationships (*see* **Note 1**). To preserve the endosymbiont's vitality, make sure the samples are as fresh as possible [3, 4].
2. *Surface Sterilization*: To get rid of any outside pollutants, surface sterilize the root samples (*see* **Note 2**). This usually includes cleaning the roots with sterile water, treating them with a number of disinfectants (such as bleach or ethanol), and then rinsing them again with sterile water [3, 4].
3. *Root Fragmentation*: To facilitate access to the endosymbionts and increase surface area, cut the sterilized roots into small segments (*see* **Note 3**).
4. *Mechanical Disruption*: To extricate the endosymbionts from the mycorrhizal tissues, mechanically disrupt the root cells (for example, by grinding or macerating them) (*see* **Note 4**).

5. *Enzymatic Digestion*: Use enzymes like cellulase and pectinase to break down the plant cell walls in the fragmented and disturbed root samples, therefore releasing the endosymbionts (*see* **Note 5**).
6. *Centrifugation and Filtration*: To get rid of bigger plant debris, filter the homogenate. To concentrate the endosymbionts, centrifuge the filtered homogenate (*see* **Note 6**).
7. *Density Gradient Centrifugation*: Utilize density gradient centrifugation to separate endosymbionts from other cellular components according to their density (*see* **Note 7**) (e.g., using sucrose or Percoll gradients) [3, 4].
8. *Isolation and Culturing*: To create pure cultures, plate the concentrated endosymbionts on appropriate growth media. Other microbes' growth can be inhibited by the use of antibiotics or selective media (*see* **Note 8**).
9. *Confirmation*: Determine the identity of the isolated endosymbionts by using molecular methods like microscopy, PCR, or sequencing (*see* **Note 9**).

Considerations

- Aseptic Techniques: To avoid contamination, aseptic conditions must be maintained throughout the procedure.
- Appropriate Media: The culture of particular endosymbionts depends on the selection of the proper growth media.
- Employ molecular techniques, such as 16S rRNA sequencing, to verify the identity of the isolated endosymbionts.

Difficulties

- Contamination: Because root and soil habitats are so complex, avoiding contamination is a major difficulty.
- Viability: It can be challenging to guarantee the endosymbionts' survival during the isolation procedure.
- Specificity: Care must be taken to isolate particular endosymbionts during the filtration and centrifugation processes without losing them [3, 4].

3.2 Detection and Identification of Bacterial Endosymbionts in Arbuscular Mycorrhizal Fungi

Detecting and identifying bacterial endosymbionts in arbuscular mycorrhizal fungi (AMF) involves a combination of molecular, microscopic, and culturing techniques. Here is a detailed outline of the process [5]:

1. *Sample Collection and Preparation*

Collect root and soil samples from plants known to form AMF associations.

Surface sterilize root samples to remove external microorganisms.

2. *DNA Extraction*

Extract DNA from the AMF-infected root samples or directly from isolated spores of AMF.

Use commercial DNA extraction kits or standard protocols involving cell lysis, purification, and isolation of DNA.

3. *PCR Amplification*

Use polymerase chain reaction (PCR) to amplify bacterial 16S rRNA genes from the extracted DNA (*see* **Note 10**).

Employ primers specific to bacterial 16S rRNA genes to avoid amplification of fungal DNA.

4. *Sequencing*

Sequence the PCR products using Sanger sequencing or next-generation sequencing (NGS) methods.

Analyze the sequences using bioinformatics tools to identify the bacterial species.

5. *Microscopic Analysis*

Fluorescence In Situ Hybridization (FISH): Use fluorescently labeled probes specific to bacterial 16S rRNA to visualize endosymbionts within the fungal structures under a fluorescence microscope (*see* **Note 11**).

Confocal Laser Scanning Microscopy (CLSM): Employ CLSM to obtain detailed images of bacterial cells within the fungal hyphae or spores.

Transmission Electron Microscopy (TEM): Use TEM to observe the ultrastructure of bacterial endosymbionts within the AMF cells.

6. *Culturing Techniques*

Attempt to culture bacterial endosymbionts from surface-sterilized AMF spores on selective media.

Use specific growth conditions and media tailored to the expected endosymbionts.

7. *Metagenomics*

Perform metagenomic analysis by sequencing the total DNA extracted from AMF samples without prior amplification.

Use bioinformatics tools to assemble and annotate the bacterial genomes, identifying the endosymbionts.

8. *Metatranscriptomics*

Extract RNA from AMF samples and convert it to cDNA.

Sequence the cDNA to study the active bacterial community and their functional roles within the AMF.

9. *Phylogenetic Analysis*

Construct phylogenetic trees using the 16S rRNA sequences to determine the evolutionary relationships of the bacterial endosymbionts.

Compare sequences with known bacterial databases to identify closely related species.

10. *Functional Characterization*

Investigate the metabolic and functional roles of the endosymbionts using techniques like.

Proteomics: Analyze the protein expression profiles of the endosymbionts.

Metabolomics: Study the metabolites produced by the endosymbionts.

Gene Expression Analysis: Use qPCR or RNA-Seq to study the expression of specific genes in the endosymbionts (*see* **Note 12**).

Considerations

- Aseptic Techniques: Ensuring all procedures are carried out under sterile conditions to avoid contamination.
- Primer Specificity: Using primers that are specific to bacterial 16S rRNA and do not amplify fungal DNA.
- Controls: Including negative and positive controls in PCR and sequencing experiments to validate results.

Challenges

- Low Abundance: Bacterial endosymbionts might be present in low numbers, making detection difficult.
- Complexity: The presence of diverse microbial communities in soil and root samples complicates the identification of specific endosymbionts.
- Viability: Culturing bacterial endosymbionts can be challenging due to specific growth requirements.

3.3 Preservation and Maintenance of Endosymbionts

The preservation of ECM fungi is quite challenging due to their inherent traits [6]. In vitro, these fungi do not produce conidia, and sexual structures are seldom formed [6] [7]. Their predominant form in vitro, hyphae, typically grows slowly in pure cultures and is generally more sensitive to environmental conditions than spores and conidia [6] [8].

The most common method for maintaining and propagating ECM fungi is in vitro sub-cultivation on suitable synthetic media, such as modified Fries medium (MFM, Colpaert et al. 2000) and modified Melin–Norkrans (MMN, Marx 1969), followed by storage at approximately 25 °C [9]. Lowering the storage temperature can lengthen the interval between subcultures; for example, Tibbett and his co-researchers [10] stored a diverse range of *Hebeloma* isolates for 3 years at 2 °C [11]. While sub-cultivation is effective for short-term preservation, it is both expensive and time-consuming [12]. Additionally, it may reduce the fungi's ability to form mycorrhizae [11] and impact essential properties, such as their capacity to enhance plant growth [11]. Regular revitalization through passage on a suitable host plant every 4 years has been also suggested [11].

3.3.1 Maintenance and Preservation of Ectomycorrhizal (ECM) Fungi

The maintenance and preservation of ectomycorrhizal (ECM) fungi involve various methods, each with specific advantages and limitations. Here are some commonly used techniques:

1. *In Vitro Sub-Cultivation*

ECM fungi are cultured on synthetic media such as modified Fries medium (MFM) and modified Melin–Norkrans (MMN). Store them typically at around 25 °C. For example, Tibbett and his co-researchers [10] stored a diverse range of *Hebeloma* isolates for 3 years at 2 °C [11].

Pros: Effective for short-term preservation.

Cons: Costly, time-consuming, and may reduce the fungi's ability to form mycorrhiza and enhance plant growth [11].

2. *Cryopreservation (Fig. 1)*

Involves freezing ECM fungi at very low temperatures, often in liquid nitrogen.

Pros: Long-term preservation, minimal genetic and physiological changes.

Cons: Requires specialized equipment and expertise.

3. *Lyophilization (Freeze-Drying)*

ECM fungi are frozen and then dried under vacuum to remove water content.

Pros: Suitable for long-term storage, easy to transport and store.

Cons: Not all fungal species survive the process.

4. *Oil Overlay Storage*

ECM fungi cultures are covered with sterile mineral oil.

Pros: Extends the shelf life of cultures without frequent sub-culturing.

Cons: Limited to certain fungal species.

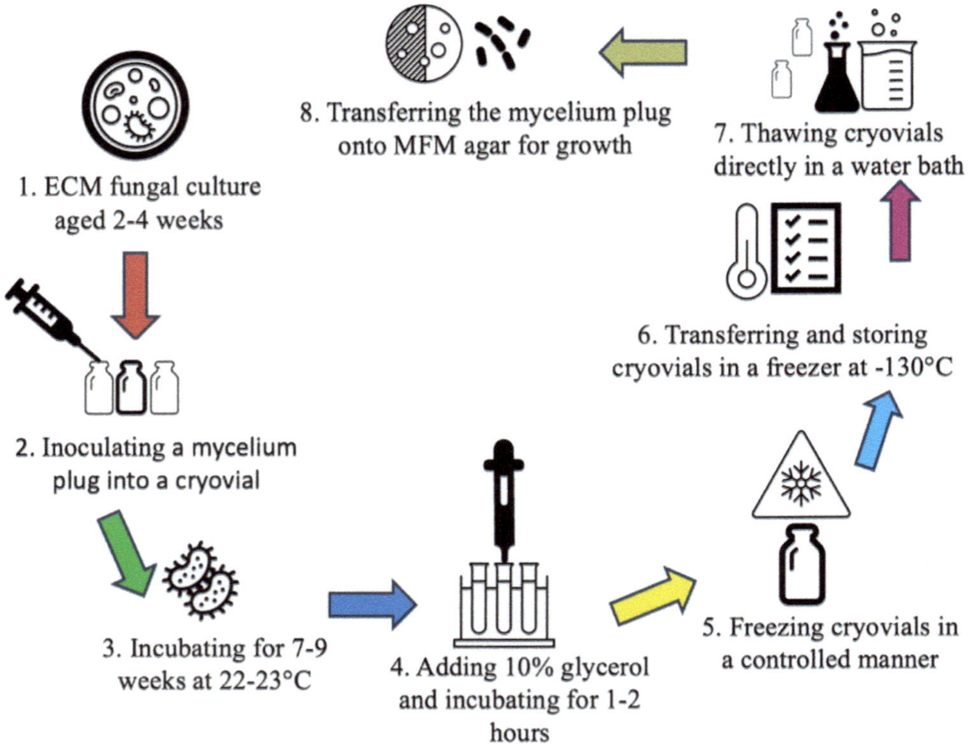

Fig. 1 Steps involved in cryopreservation of ECM fungi

5. *Storage on Host Plants*

Periodic transfer of ECM fungi to suitable host plants to maintain their vitality. Recommended every 4 years [11].

Pros: Maintains the fungi's symbiotic capabilities and plant growth-promoting properties.

Cons: More labor-intensive and requires access to host plants.

6. *Storage at Reduced Temperatures*

Storing cultures at lower temperatures (e.g., 2 °C) to slow down metabolic processes.

Pros: Extends the period between subcultures.

Cons: May still require periodic sub-culturing and monitoring.

7. *Use of Carriers and Substrates*

ECM fungi are inoculated onto carriers such as perlite, vermiculite, or soil substrates.

Pros: Mimics natural conditions, can be used directly for inoculation in the field.

Cons: May not be suitable for long-term storage without additional preservation methods.

Maintaining and preserving arbuscular mycorrhizal (AM) fungi are crucial for research and agricultural applications. These fungi form symbiotic relationships with plant roots, aiding in nutrient uptake and improving plant health. Here are the main methods used for the maintenance and preservation of AM fungi.

3.3.2 Maintenance and Preservation of AM fungi.

1. *In Vivo Culture*

AM fungi are cultured with living host plants, typically in pots with sterilized soil or substrate.

Pros: Mimics natural conditions, ensuring fungi maintain their symbiotic properties.

Cons: Space-intensive, labor-intensive, and may not be suitable for long-term storage.

2. *In Vitro Culture*

AM fungi are grown with transformed root organ cultures, such as carrot roots, in a controlled laboratory environment.

Pros: Allows for controlled studies and easier observation, requires less space than in vivo culture.

Cons: May not fully replicate the complex interactions found in natural soil environments.

3. *Cryopreservation*

AM fungi are stored at very low temperatures, typically in liquid nitrogen.

Pros: Suitable for long-term storage, minimal genetic and physiological changes.

Cons: Requires specialized equipment and expertise.

4. *Cold Storage*

Storing AM fungi at low temperatures, usually just above freezing, to slow down metabolic processes.

Pros: Extends the period between sub-culturing, simpler than cryopreservation.

Cons: Limited storage duration compared to cryopreservation.

5. *Air-Drying and Storage in Dry Conditions*

Spores of AM fungi are air-dried and stored in dry, cool conditions.

Pros: Simple and cost-effective method for short- to medium-term storage.

Cons: Not suitable for long-term preservation, may reduce viability over time.

6. *Encapsulation in Alginate Beads*

AM fungi spores or mycelia are encapsulated in alginate beads and stored.

Pros: Protects the fungi and allows for easy handling and transport.

Cons: Requires specific preparation methods and may not be suitable for all fungal strains.

7. *Use of Substrates and Carriers*

AM fungi are inoculated onto substrates like vermiculite, perlite, or soil and stored.

Pros: Mimics natural growth conditions, suitable for large-scale applications.

Cons: Storage duration may be limited, and substrates can degrade over time.

8. *Maintenance through Regular Sub-Culturing*

Regular transfer of AM fungi to fresh host plants or new culture media to maintain viability.

Pros: Ensures ongoing vitality and symbiotic capabilities.

Cons: Labor-intensive and requires regular monitoring and resources.

9. *Lyophilization (Freeze-Drying)*

AM fungi are frozen and dried under vacuum to remove water content.

Pros: Suitable for long-term storage, easy to transport.

Cons: Not all AM fungi survive the process, and it requires specialized equipment.

10. *DNA Storage*

DNA of AM fungi is extracted and stored for future use.

Pros: Allows for genetic studies and long-term preservation of genetic material.

Cons: Does not preserve the living organism or its symbiotic properties.

3.4 Future Directions for Isolation and Maintenance of ECM and AM Fungi

The isolation and maintenance of ectomycorrhizal (ECM) fungi and arbuscular mycorrhizal (AM) fungi are pivotal for advancing ecological research and agricultural practices. Here are some future directions and trends in this field [13]:

1. *Advanced Molecular Techniques*

Genomic and Metagenomic Approaches: Enhanced sequencing technologies will allow for deeper insights into the genetic diversity and functionality of mycorrhizal fungi. This can aid in identifying specific traits that enhance symbiosis and stress tolerance.

CRISPR and Genetic Engineering: Applying CRISPR and other gene-editing technologies could enable precise modifications to improve fungal traits, such as increased nutrient exchange efficiency or stress resistance.

2. *Improved In Vitro Culture Systems*

Synthetic Microenvironments: Development of more complex synthetic media that better mimic the natural soil environment, including specific chemical and physical properties, to support the growth and maintenance of both ECM and AM fungi.

Co-Culture Systems: Creating co-culture systems that include multiple species of fungi, bacteria, and host plants to simulate natural symbiotic networks is more accurate.

3. *Cryopreservation Innovations*

Optimized Protocols: Refinement of cryopreservation techniques to improve survival rates and viability of ECM and AM fungi. This includes optimizing cryoprotectants and freezing/thawing protocols.

Automated Cryopreservation: Development of automated systems for large-scale cryopreservation and storage, facilitating easier and more efficient long-term preservation.

4. *Bioinformatics and Data Integration*

Big Data Analytics: Leveraging big data analytics to analyze large datasets from sequencing, metabolomics, and proteomics studies, providing comprehensive insights into mycorrhizal functions and interactions [14].

Integrated Databases: Creating integrated databases that combine genetic, physiological, and ecological data on ECM and AM fungi, enhancing accessibility and utility for researchers.

5. *Ecological and Environmental Studies*

Field-Based Monitoring: Enhanced field-based monitoring using sensors and remote sensing technologies to study mycorrhizal networks in their natural habitats, providing real-time data on their dynamics and responses to environmental changes.

Climate Change Adaptation: Research focused on understanding how ECM and AM fungi can help plants adapt to climate change, including drought tolerance, temperature extremes, and soil degradation.

6. *Sustainable Agricultural Practices*

Bioinoculants Development: Developing efficient and scalable methods for producing mycorrhizal bioinoculants that can be used in sustainable agriculture to enhance crop yield and resilience.

Customized Inoculants: Tailoring mycorrhizal inoculants to specific crops and environmental conditions, improving their effectiveness and compatibility.

7. *Integration with Host Plants*

Host–Plant Genetics: Studying the genetic factors in host plants that influence mycorrhizal associations, potentially leading to the breeding of crop varieties with enhanced symbiotic capabilities.

Symbiosis Optimization: Investigating and optimizing the signaling pathways and molecular interactions between fungi and host plants to strengthen and sustain symbiotic relationships.

8. *Public and Farmer Education*

Outreach Programs: Developing educational programs for farmers and the public to raise awareness about the benefits of mycorrhizal fungi in agriculture and ecosystems.

Collaborative Research: Encouraging collaborative research efforts between academic institutions, agricultural organizations, and industry to drive innovation and application of mycorrhizal technologies.

By focusing on these future directions, researchers and practitioners can enhance the isolation, maintenance, and application of ECM and AM fungi, leading to more sustainable and resilient ecosystems and agricultural systems.

4 Notes

1. Roots from known mycorrhizal plants are chosen to ensure the presence of symbiotic fungi. Fresh samples maintain the viability of the endosymbionts for successful isolation. Delayed processing can result in fungal death or contamination. Field-to-lab time should be minimized to preserve microbial integrity.
2. Sterilization removes external contaminants like bacteria or fungi from the root surface. Disinfectants such as bleach or ethanol are used in controlled concentrations. This step ensures that only internal endosymbionts are studied. Multiple sterile water rinses help remove residual disinfectants.
3. Sterilized roots are cut into small pieces to increase surface area. This facilitates enzyme and mechanical access to internal tissues. Smaller segments also make homogenization more efficient. Uniform fragments improve consistency in downstream processing.
4. Physical methods like grinding release endosymbionts from root tissues. This breaks open plant cells while trying to preserve fungal structures. Disruption must be thorough but not overly harsh to avoid damaging symbionts. It prepares the sample for more refined separation.

5. Enzymes like cellulase and pectinase digest plant cell walls. This softens root tissues and helps release internal microorganisms. It complements mechanical methods for thorough extraction. Controlled enzyme use prevents damage to fungal cells.
6. Filtration removes large debris, leaving behind a microbial-rich solution. Centrifugation concentrates the endosymbionts by spinning down heavier particles. This step purifies the sample before density separation. Proper speed and time are critical for optimal yield.
7. A gradient (e.g., sucrose or Percoll) separates components by density. Endosymbionts settle at specific layers, allowing targeted extraction. This enhances purity by removing unwanted cell fragments. It is a key step for isolating viable, concentrated fungi.
8. The purified fungi are plated on selective media to grow pure colonies. Antibiotics may suppress bacterial contaminants. Different endosymbionts require different culture conditions. Growth media should match the fungi's nutritional and environmental needs.
9. Microscopy, PCR, or sequencing confirms the identity of the isolated fungi. Molecular techniques like 16S/ITS sequencing provide accurate classification. This ensures the right organism has been isolated and cultured. It is essential for downstream research or application.
10. PCR is used to amplify bacterial 16S rRNA genes because this region is highly conserved and ideal for identifying bacteria. Amplification makes it easier to detect and analyze even small amounts of bacterial DNA. The 16S rRNA gene provides species-level resolution through sequencing and comparison with known databases.
11. *Fluorescence in situ hybridization* is a molecular technique used to detect and locate specific DNA or RNA sequences within cells or tissues. It uses fluorescently labeled probes that bind to target sequences, allowing visualization under a fluorescence microscope. FISH is widely used to identify microbial species, detect chromosomal abnormalities, or study gene expression at the cellular level.
12. The functional characterization of endosymbionts is essential to understanding their biological roles and interactions with host organisms. Analyzing protein expression profiles reveals the active biological processes and metabolic pathways in endosymbionts. It helps identify proteins involved in symbiosis, nutrient exchange, or stress responses. This information provides direct insight into the functional behavior of the endosymbionts at the molecular level. Studying metabolites reveals the chemical compounds produced or modified by

endosymbionts. This uncovers their roles in nutrient provisioning, defense, or signaling within the host. It bridges the gap between gene expression and observable phenotypic effects. Gene expression profiling identifies which genes are up- or downregulated under specific conditions. It reveals how endosymbionts respond to environmental or host-related stimuli. Techniques like qPCR and RNA-Seq offer quantitative and comprehensive data, supporting functional gene annotation and regulatory studies. Together, these approaches provide a *multilayered understanding* of endosymbiont function—linking genotype to phenotype and molecular activity to ecological roles.

References

1. Tzng E (2022) Studying Arbuscular Mycorrhizal fungi, Their Endosymbiotic Bacteria, and Their Spores: Challenges and Solutions
2. Guillén A (2021) Preparation of samples for characterization of arbuscular mycorrhizal fungi. In: The plant microbiome: methods and protocols, pp 43–51
3. Bonfante P (2003) Plants, mycorrhizal fungi and endobacteria: a dialog among cells and genomes. Biol Bull 204(2):215–220
4. Kyde MM, Gould AB (2000) Mycorrhizal endosymbiosis. In: Microbial endophytes. CRC Press, pp 175–212
5. Bianciotto V, Lumini E, Lanfranco L, Minerdi D, Bonfante P, Perotto S (2000) Detection and identification of bacterial endosymbionts in arbuscular mycorrhizal fungi belonging to the family Gigasporaceae. Appl Environ Microbiol 66(10):4503–4509
6. Pellitier PT, Zak DR (2018) Ectomycorrhizal fungi and the enzymatic liberation of nitrogen from soil organic matter: why evolutionary history matters. New Phytol 217(1):68–73
7. Dyer PS, O'Gorman CM (2012) Sexual development and cryptic sexuality in fungi: insights from *Aspergillus* species. FEMS Microbiol Rev 36(1):165–192
8. Senanayake IC, Rathnayaka AR, Marasinghe DS, Calabon MS, Gentekaki E, Lee HB, Xiang MM (2020) Morphological approaches in studying fungi: collection, examination, isolation, sporulation and preservation. Mycosphere 11(1):2678–2754
9. Hinman SS, Wang Y, Kim R, Allbritton NL (2021) *In vitro* generation of self-renewing human intestinal epithelia over planar and shaped collagen hydrogels. Nat Protoc 16(1): 352–382
10. Tibbett M, Sanders FE, Cairney JWG (1999) Long-term storage of ectomycorrhizal basidiomycetes (Hebeloma spp.) at low temperature. J Basic Microbiol 39(5–6):381–384
11. Lalaymia I, Cranenbrouck S, Declerck S (2014) Maintenance and preservation of ectomycorrhizal and arbuscular mycorrhizal fungi. Mycorrhiza 24:323–337
12. Vinzelj J, Joshi A, Young D, Begovic L, Peer N, Mosberger L, Podmirseg SM (2022) No time to die: comparative study on preservation protocols for anaerobic fungi. Front Microbiol 13: 978028
13. Chu W, Guo X, Zhang C, Zhou L, Wu Z, Lin W (2022) Research progress and future directions of arbuscular mycorrhizal fungi-plant-rhizosphere microbial interaction. Chin J Eco-Agric 30(11):1709–1721
14. Wijayawardene NN, Boonyuen N, Ranaweera CB, de Zoysa HK, Padmathilake RE, Nifla F, Chen HH (2023) OMICS and other advanced technologies in mycological applications. J Fungi 9(6):688

Chapter 11

Isolation and Maintenance of Endosymbionts from Lichens

Rajvi Mehta and Natarajan Amaresan

Abstract

Lichens are made due to a symbiotic relationship between a mycobiont and a photobiont. Lichens are a model system to study symbiosis. However, the challenges in the isolation of lichen endosymbionts contribute to the difficulty in the area of research in lichen. This chapter aims to describe various methods used for the isolation and maintenance of lichen endosymbionts.

Key words Endosymbionts, Lichen isolation, Photobionts, Mycobionts, Micropipette method, Spray method

1 Introduction

Lichens represent symbiotic interactions of organisms between a fungus and a photobiont (algae or cyanobacterium). The relationship between them is complex where lichens provide bacteria robust microenvironments for bacterial colonization. Lichen-associated bacteria are an essential component of the lichen microsystem [1]. Lichens are made up of a compact structure of fungal hyphae that are occupied by different types of bacteria [2]. Endosymbionts denote photobionts (photosynthetic member) and mycobionts (fungal member) in the lichen system. Lichens are capable of colonizing in extreme environments which contributes to global biodiversity. The study of lichen endosymbionts provides a distinct perspective on symbiotic relationships and the development of complex organisms. However, isolation of lichen symbionts is challenging because lichens have a compact structure and the nature of lichen endosymbionts is delicate. Moreover, the risk of contamination in isolated cultures is high due to lichens harboring numerous microorganisms. These reasons make the isolation of lichen endosymbionts difficult and complex, but still, various studies have developed methods to successfully isolate symbionts from lichens. Therefore, this chapter depicts some of the

N. Amaresan and Sapna Chandwani (eds.), *Endosymbionts*, Springer Protocols Handbooks,
https://doi.org/10.1007/978-1-0716-4771-4_11,

most common methods that are used for the isolation and maintenance of photobionts and mycobionts from lichens. The procedure for isolation of mycobionts contains isolation from spores and direct isolation from the thallus segment. The methods for isolation of photobionts described in this chapter are the spray method and the micropipette method. Methods for maintenance of cultured endosymbionts are also mentioned by Weidman et al. [3] and Yoshimura et al. [4].

2 Materials

2.1 Equipment

The types of equipment needed in the isolation procedure of endosymbionts are as follows:

Compound microscope

Inverted microscope

Ultra sonicator

Centrifuge

Autoclave

Laminar airflow

Capillary tube

Sterile micropipettes (*see* **Note 1**)

2.2 Culture Media for Mycobionts and Photobionts

Water agar medium and malt/yeast extract medium are commonly used media for spore collection and germination [5, 6].

1. Distilled water agar medium (4%)

Agar	4 g
Distilled water	100 mL

2. Malt/yeast extract medium

Malt extract	20 g
Yeast extract	2 g
Agar	20 g
Distilled water	1000 mL

3. Lilly and Bernett's medium

Glucose	10.0 g
Asparagine	2.0 g

(continued)

KH_2PO_4	1.0 g
$MgSO.7H_2O$	0.5 g
$Fe(NO_3)_3.9H_2O$	0.2 mg
$MnSO_4.4H_2O$	0.1 mg
$ZnSO_4.7H_2O$	0.2 mg
Thiamine	0.1 mg
Biotin	5 μg
Distilled water	1000 mL

Lilly and Bernett's medium [7] is also used for isolation of mycobionts. Yamamoto et al. [8] created Lilly and Bernett's Gelrite medium by modifying Lilly and Bernett's medium by using 1% w/v gel rite instead of agar.

2.3 Culture Media for Photobionts

Bold's basal medium (BBM), Trebouxia organic nutrient medium, and medium differential medium (MDM) medium are generally used mediums for isolating photobionts [5, 9, 10]

Bold's basal medium (BBM, 1000 mL)

$NaNO_3$	250 mg
KH_2PO_4	175 mg
K_2HPO_4	75 mg
$MgSO_4.7H_2O$	75 mg
$CaCl_2.2H_2O$	25 mg
NaCl	25 mg
EDTA	50 mg
KOH	31 mg
H_3BO_3	11.42 mg
$ZnSO_4.7H_2O$	8.82 mg
$FeSO_4.7H_2O$	4.98 mg
$MnCl_2.7H_2O$	1.44 mg
MoO_3	0.71 mg
$CuSO_4.5H_2O$	1.57 mg
$Co(NO_3)_2.6H_2O$	0.49 mg

BBM agar medium can also be used (*see* **Note 1**).

Trebouxia organic nutrient medium

BBM	970 mL
Protease peptone	10 g
Glucose	20 g

MDM medium Watanabe et al. [10]

KNO_3	1 g
$MgSO_4.7H_2O$	250 mg
NaCl	250 mg
K_2HPO_4	100 mg
$CaCl_2.2H_2O$	10 mg
Fe solution	1 mL
A5 solution	1 mL
Distilled water	1000 mL

Fe solution

$FeSO_4$	1 g
Distilled water	500 mL
Concentrated H_2SO_4	2 drops

A5 solution

H_3BO_3	286 mg
$MnSO_4.7H_2O$	250 mg
$ZnSO_4.7H_2O$	22.2 mg
$CuSO_4.7H_2O$	7.9 mg
Na_2MoO_4	2.1 mg
Distilled water	100 mL

3 Methods

3.1 Pretreatment of Sample for Isolation of Endosymbionts

Take the thallus of lichen and cut off approximately 1 cm^2 from the apex. Continuously rinse it for 5–10 min using tap water. Clean the thallus surface by brushing it with a paintbrush in running water. Wash the thallus again with sterilized water.

For small lichen thallus, take a small test tube filled with 1–2 mL sterile distilled water and one drop of Tween 20 and then add the thallus to the tube. Ultra-sonicate the suspension for 3 min. For the final step, centrifuge the suspension at 2000 rpm.

3.2 Separation of Symbionts

1. Take a sterilized glass microscopic slide and mount it with lichen thallus.
2. Then, scrape the thallus surface by using a small needle under a microscope.
3. Separate the photobiont layer from the mycobiont layer and move it to a sterilized glass slide. Grind the photobiont layer by adding a drop of sterile distilled water and then covering it with another glass slide. This will create small fragments of the photobiont layer.
4. Now suspend both mycobiont and photobiont in the liquid. *See* **Note 2** for the microlichen separation method.
5. For isolation, take BBM or MDM culture agar plate and inoculate the endosymbiont culture solution by adding one drop on the agar plate. Incubate the plates at 15 °C under the light intensity of 10–27 μmol m^{-2} s^{-1} for 25–30 days.
6. Observe the appearance of colonies on the agar surface after the incubation of 25–30 days.
7. For confirmation of true photobiont, check the fungal tissue–associated cell.
8. To retain only the original photobiont and eliminate any contamination, observe colonies without contamination under a stereomicroscope and transfer the colony onto a suitable agar medium. Generally, the Trebouxia organic medium is used for green algae, and the MDM medium supplemented with 1% glucose is used for cyanobacteria.

3.3 Micropipette Method

This method was described by Yoshimura et al. [11] for isolation of lichen endosymbionts.

1. Take a glass microscopic slide containing wells and add a drop of sterile water/sterile culture media to the wells. Place the algal colony on the well.
2. Pick up about five to ten algal cells using the micropipette (*see* **Note 3**).
3. Blow the cells through the glass tubing of the micropipette and erupt the cells into the next well. This technique should be repeated more than five times or until the contamination is removed.
4. Transfer around 10–50 algal cells in a new culture medium (transfer one single cell at a time).

5. To ensure successful isolation, culture the algal cells in the Trebouxia organic nutrient medium and MDM media containing glucose. For complete purification of cyanobacteria, use micromanipulation methods.

3.4 Isolation of Photobionts Using Spray Method

This method was used by Weidman et al. [3] for the isolation of a single cell of algae without any contamination.

1. Select colonies from the agar plates, and transfer them to a test tube containing 1x N BBM slant agar medium. Let the colonies grow for several weeks.
2. Take sterile centrifugation tubes and add 1 mL sterilized water and one drop of Tween 20 in the tubes. Transfer the colonies to the tubes.
3. Ultrasonicate the culture. Observe the dissociation of photobiont colonies from other colonies and the attachment of mycobionts on the surface of cell walls.
4. Centrifuge the ultrasonicated culture at 1000 rpm for 5–10 min. Discard the supernatant and add 1 mL sterile water and one drop of Tween 20 in the centrifuge tube containing photobiont cells (*see* **Note 3**).
5. Take the centrifuge tube and insert a capillary tube at the bottom. To draw algal suspension in the capillary tube, directly blow the air through the small opening of the capillary tube. The algal suspension will be atomized in a spray form.
6. Spray a Petri dish consisting of a suitable medium so that algal cell suspension can be coated onto the Petri plate.
7. Let the culture grow for about 1–2 weeks and observe for non-contaminated colonies. Transfer these colonies to a freshly prepared culture medium.

3.5 Isolation of Mycobionts

Mycobionts are usually isolated from spores. Yoshimura et al. [11] gave the following procedure for the isolation of mycobionts.

1. Wash the thalli that are sampled from the field properly. Take the thallus and separate the apothecia and perithecia. Soak them in sterile distilled water for about 4 hours. Remove excess water from the spore-bearing material by blot drying. Next, take a plastic Petri dish and fix the above structures to the bottom with the help of petroleum jelly. Cover the top layer by adding 4% water agar medium to reduce contamination.
2. The agar placed on the upper lid should be in the 5–10 mm range of the lower lid. Set the discharge time and distance between two lids according to single-spore or multiple-spore isolation.

3. Change the Petri plate with a plate containing fresh medium from time to time. For germination, cover the Petri plates by sealing them with parafilm and incubate at 15 °C in dark condition.
4. Observe the germination process by checking spores released on the water agar medium using an inverted microscope. Transfer the spores on a microscope well slide consisting of agar medium. Maintain the slides under humid condition and monitor germination.

3.6 Isolation of Mycobionts from the Thallus

When there is no apothecia to isolate mycobionts, thallus fragments can be used to isolate mycobionts.

1. Take a fresh thallus and cut it into small segments with the help of a sterilized razor. Store it in test tubes containing only water or wet filter paper at 15 °C in humid condition. Monitor the tubes for growth of medullary hyphae. Remove a section of elongated hyphae using the aseptic method. Transfer the section to a freshly prepared culture medium (*see* **Note 4**).
2. To confirm the growth of mycobiont, repeat the above procedure in several replicas.

3.7 Maintenance of the Endosymbiont Cultures

Generally, storage period for mycobiont cultures is 1 year. Although, subculturing the mycobiont culture every 2–3 months is recommended. To subculture, make various segments of mycobiont colonies by cutting them with a scalpel. Culture the segments on malt/yeast extract agar (MY agar) or LB agar medium at 15 °C for 2–3 months. Maintain optimum temperature (15–20 °C) and pH (*see* **Note 5**) Yoshimura et al. [11].

For photobiont cultures, subculturing should be done every 2–3 months. When selecting culture conditions, the temperature at which the lichen grows naturally should be considered. The pH should be maintained at 4.0–7.0.

If there is a need for storage of mycobiont culture for a long period of time, cryopreservation can be done.

4 Notes

1. Add 15–20 gm agar in the listed components to make a solid medium.
2. Use the same method for microlichens but apply higher pressure on the thallus (cortex). Both photobiont and mycobiont will be suspended after this step. Use a mortar and pestle or blender to break up bigger pieces into smaller ones.
3. For preparation of a special micropipette, heat a glass tube having a 4–5 mm diameter at the center. After heating, stretch

the tube from the ends and divide the tube at the center. At the wider end, place a cotton stopper and connect a rubber tube.

4. The procedure should be repeated at least 20 times for proper isolation without contamination.
5. Optimum temperature and pH vary according to species. Prior study of the species is essential for that matter.

References

1. Lee YM, Kim EH, Lee HK, Hong SG (2014) Biodiversity and physiological characteristics of Antarctic and Arctic lichens-associated bacteria. World J Microbiol Biotechnol 30:2711–2721
2. Erlacher A, Cernava T, Cardinale M, Soh J, Sensen CW, Grube M, Berg G (2015) Rhizobiales as functional and endosymbiotic members in the lichen symbiosis of Lobaria pulmonaria L. Front Microbiol 6:53
3. Weidman VE, Walne P, Tainor F (1984) A new technique for obtaining axenic culture of algae. Can J Bot 42(7):958–959
4. Yoshimura I (2002) Isolation and culture of lichen photobionts and mycobionts. In: Protocols in lichenology: culturing, biochemistry, ecophysiology and use in biomonitoring. Springer, pp 3–33
5. Ahmadjian V (1967) A guide to the algae occurring as lichen symbionts: isolation, culture, cultural physiology, and identification. Phycologia 6(2–3):127–160
6. Pyatt FB (1973) Lichen propagules. In: The lichens, pp 117–145
7. Lilly VG, Barnett HL (1951) Physiology of the fungi, vol 12. McGraw-Hill Book Company Inc., p 464
8. Yamamoto Y, Kinoshita Y, Takahagi T, Kroken S, Kurokawa T, Yoshimura I (1998) Factors affecting discharge and germination of lichen ascospores. J Hattori Bot Lab 85:267–278
9. Brown RM, Bold HC (1964) Phycological studies V. Comparative studies of the algal genera *Tetracystis* and *Chlorococcum*. The University of Texas Publication, p 6417
10. Watanabe A (1960) List of algal strains in collection at the Institute of Applied Microbiology, University of Tokyo. World J Microbiol Biotechnol 6(4):283–292
11. Yoshimura I, Kurokawa T, Kanda H (1990) Tissue culture of some antarctic lichens preserved in the refrigerator. Proc NIPR Symp Polar Biol 3:224–228

Part II

Methods for Identification of Endosymbionts

Chapter 12

Microscopy and Staining-Based Identification of Endosymbionts

Sephokoane C. Makuwa and Mahloro H. Serepa-Dlamini

Abstract

Endosymbionts such as bacteria and fungi play crucial roles in the survival and growth of host plants. The identification of these microorganisms often requires a combination of traditional staining methods, such as Gram staining, and the use of complex microscopy such as scanning electron microscope. Microscopic techniques, including light, fluorescence, confocal, and electron microscopy, provide insights into endosymbionts' morphology, localization, and structure within the host tissues. This chapter outlines the protocols for the morphological examination and microscopic visualization of endosymbionts.

Key words Endosymbionts, Gram staining, Confocal microscopy, Electron microscopy, Bacterial morphology, Molecular identification, Scanning electron microscope, Endosymbionts

1 Introduction

Endosymbiont microbes are typically bacteria or fungal species that reside and thrive inside tissues of plant hosts [1]. The relationship is symbiotic, and both parties benefit each other. The plant provides an environment for the endosymbiont to survive and nutrients such as carbohydrates, lipids, amino acids, and vitamins [2]. In contrast, the endosymbiont promotes plant growth through phytohormone production, nutrient acquisition, and defense against herbivores and microbial pathogens by the production of antimicrobial compounds. Endosymbionts can be exploited for biotechnological applications, including improving crop yields, phytoremediation, and producing bioactive compounds with a variety of applications; thus, their isolation, identification, and characterization are of importance [2, 3]. A variety of staining and microscopy techniques are employed to study endosymbionts, as these methods are invaluable for their identification, localization, and characterization within plant tissues. These approaches provide key insights into the structure, spatial distribution, and interactions of

N. Amaresan and Sapna Chandwani (eds.), *Endosymbionts*, Springer Protocols Handbooks,
https://doi.org/10.1007/978-1-0716-4771-4_12,

endosymbionts with their hosts, particularly when molecular techniques alone are insufficient [4, 5]. By combining microscopy techniques, researchers can identify these microorganisms and visualize their structures. This chapter details protocols for the morphological identification of bacterial endosymbionts, focusing on microscopy and staining-based techniques.

2 Materials

2.1 Plant Sample Collection and Isolation of Endosymbionts

1. Sterile polyethylene bags.
2. Portable cooler with ice packs.
3. 70% ethanol.
4. Sterile distilled water.
5. 2% sodium hypochlorite (NaClO).
6. Sterile mortar and pestle.
7. Phosphate buffer.
8. Incubator (set at 30 °C).
9. Glycerol (30% v/v solution).
10. −80 °C freezer.

2.2 Gram Staining

1. Clean and dry glass slides.
2. Sterilized distilled water.
3. Sterilized inoculation loop.
4. Crystal violet solution.
5. Gram's iodine solution (mordant).
6. 95% ethyl alcohol (decolorizing agent).
7. Safranin (counterstain).
8. Absorbent paper.
9. Bright-field microscope.

2.3 Fluorescent in Situ Hybridization (FISH)

1. Leaf samples from targeted plants.
2. Ethanol (80%).
3. KOH solution (10%, w/v).
4. H_2O_2 solution (30%, v/v).
5. HCl solution (10%, v/v).
6. Sterile distilled water.
7. Plate count agar (PCA).
8. Poly-L-lysine-coated slides.
9. Paraformaldehyde (4%, v/v).

10. Proteinase K (10 mg/mL).
11. EUB338 probe (labeled with fluorescein).
12. Hybridization buffer (25% formamide, 0.9 M NaCl, 0.01% SDS, 20 mM Tris–HCl (pH 7.2)).
13. Washing buffer (20 mM Tris–HCl (pH 7.2), 0.01% SDS, 40 mM NaCl, 5 mM EDTA).
14. AxioCam MRc5 camera.
15. Software: AxioVs40 V 4.6.1.0, LSM 510 version 4.2.

2.4 Fluorescence Confocal Microscopy

1. DAPI (4′,6-diamidino-2-phenylindole).
2. Paraformaldehyde (3%).
3. Cold 50 mM HEPES buffer (pH 7.5).
4. Centrifuge (500×*g*).
5. Microscope slides.
6. Plan Apo 100/1.4 oil lens.
7. Confocal scan head (LSM510 Meta).
8. DAPI staining solution (1 μg/ml in 50 mM HEPES).
9. Cy3 (fluorescence dye).
10. Cy5 (fluorescence dye).
11. Lasers: 30-mW Diode 405-nm, 1.2-mW HeNe 543-nm, 5-mW HeNe 633-nm.
12. Differential interference contrast (DIC) imaging setup.

2.5 Transmission Electron Microscopy (TEM)

1. 2.5% glutaraldehyde in 0.1 M cacodylate buffer (pH 7.3).
2. agarose gel.
3. 2% osmium tetroxide.
4. 2% uranyl acetate.
5. Ethanol (30%, 50%, 70%, 95%, 100%).
6. Propylene oxide.
7. Epon embedding resin.
8. Pioloform-coated copper grids.
9. Saturated uranyl acetate.
10. Lead citrate.
11. Leica Ultracut UCT microtome (Leica Microsystems).
12. A transmission electron microscope.

2.6 Scanning Electron Microscopy (SEM)

1. Phosphate-buffered saline (PBS).
2. 0.25% glutaraldehyde (in sodium phosphate, pH 7.2).

3. Incubation system (room temperature and overnight incubation).
4. Polycarbonate (nuclepore) filter membranes (0.45 μm or 0.2 μm).
5. 2.5% glutaraldehyde in PBS.
6. 1% osmium tetroxide (OsO_4) in PBS buffer.
7. Phosphate buffer.
8. Ethanol (30%, 50%, 70%, 80%, 90%, 95%, and 100%).
9. Centrifuge.
10. SEM stubs with adhesive tape.
11. Critical point dryer.
12. Sputter coater (gold).
13. Luria broth.
14. Incubator (30 °C, 150 rpm shaker).
15. Centrifuge (1100×*g*).
16. 1% formaldehyde and 2% glutaraldehyde (1:1, v/v).
17. Distilled water.
18. Ethanol (30%, 50%, 70%, 90%, 95%, and 100%).
19. EmScope SC 500 (gold coating system).
20. TESCAN VEGA SEM (VEGA 3 LHM, AVG9731276ZA) connected to a monitor.

3 Methods

3.1 Plant Sample Collection and Isolation of Endosymbionts

1. Collect plant samples from the specified locations (*see* **Note 1**).
2. Record the geographical coordinates that specify the exact latitude and longitude of a point on the Earth's surface. For example, Limpopo Province, South Africa (−23.492657, 29.713299), according to Jasim et al. [3].
3. Collect the whole or plant tissue samples of interest and place the samples in sterile polyethylene bags. Transport them to the laboratory while maintaining a temperature of 4 °C using ice packs.
4. Perform formal identification of the plant specimens at the herbarium, where plant herbarium specimens are also deposited.
5. Wash the collected plant parts thoroughly with running tap water to remove surface debris. Surface sterilize the plant parts by immersing them in 70% ethanol for 5 min (*see* **Note 2**).

6. Rinse the plant parts with sterile distilled water. Soak them in 2% sodium hypochlorite (NaClO) for 3 min to further disinfect the surfaces.
7. Perform three final rinses with sterile distilled water to remove residual sterilizing agents. Plate the final rinse water onto nutrient agar plates as a control for sterility.
8. Transfer the surface-sterilized plant parts to a sterile mortar and pestle. Crush and macerate the plant parts to release endophytic bacteria (*see* **Note 3**).
9. Add phosphate buffer (8 g NaCl, 0.2 g KCl, 1.44 g Na_2HPO_4, and KH_2PO_4, pH 7.4) to the macerated plant material to create a homogenate.
10. Using sterile techniques, streak the homogenate onto nutrient agar plates. Incubate the plates at 28 °C for 2–7 days and monitor daily for bacterial colony growth.
11. Subculture any bacterial colonies onto fresh nutrient agar plates to obtain pure single colonies.
12. Confirm culture purity by analyzing phenotypic characteristics, including colony morphology, color, size, and Gram staining. Select pure bacterial isolates for further study.
13. Prepare a 30% glycerol solution for long-term storage. Suspend the selected bacterial isolates in the glycerol solution, and gently mix and store the glycerol stock solutions at −80 °C for preservation.

3.2 Gram Staining of Endosymbionts

Gram staining is used to classify bacteria based on the structure of their cell walls [5, 6]. Moreover, the morphological structure, including the isolates' cell shape, size, and arrangements, can be conducted by Gram staining using pure cultures, according to Cappucino and Sherman [7].

1. Place a drop of sterilized distilled water on a clean and dry glass slide.
2. Using a sterilized inoculation loop, transfer a colony of actively growing bacterial culture to a water drop, and prepare a thin and uniform smear.
3. Air-dry the smear and gently heat fix by passing the slide through a flame (*see* **Note 4**).
4. Gently flush the bacterial smear with crystal violet for 1 min (*see* **Note 5**).
5. Gently wash the slides with tap water for 2 s.
6. Flood the slide with Gram's iodine (mordant) and wait for 1 min.
7. Gentle wash the slides with tap water for 2 s.

8. Apply the decolorizing agent (95% ethyl alcohol) for 10 s, and gently wash the slide for 2 s (*see* **Note 6**).
9. Flood the slide with safranin and wait for 1 min.
10. Gently wash the slide with tap water, and blot dry with absorbent paper.
11. Observe the stained slide under the oil-immersion objective lens using a bright-field compound microscope and then note the morphological details such as Gram stain color, cell shape, and arrangement (*see* **Note 7**).

3.3 Fluorescent In Situ Hybridization (FISH)

The FISH is a powerful tool for identifying endosymbionts by targeting their ribosomal RNA sequences with fluorescently labeled probes [8–11].

3.3.1 Foliar Sampling

1. Collect leaf samples from selected plants based on whether they are colonized by the endosymbiont or not (*see* **Note 8**).
2. Harvest two leaves per plant and cut five foliar disks (0.7 cm diameter) from each leaf.

3.3.2 Leaf Sampling

1. Select disease-free leaf samples from plants (two leaves per plant).
2. Cut five foliar disks (0.7 cm diameter) from each leaf.

3.3.3 Leaf Decolorization

1. Soak the leaf disks in 80% ethanol for 2 weeks.
2. Boil the disks in a 10% KOH solution for 20 min.
3. Transfer the disks to a 30% H_2O_2 solution until they get completely cleared (20–40 min).
4. Acidify the samples by quick immersion in 10% HCl and rinse with sterile distilled water.
5. Place the samples in sterile distilled water for 15 days before FISH analysis.

3.3.4 Viable Bacterial Count

1. Crush one leaf disk and plate it on plate count agar (PCA) to isolate endophytic bacteria.
2. Place one disk on the PCA for further analysis.
3. Incubate the plates at 25 °C for 3 days and purify the isolates by repeated streaking on PCA.
4. Compare viable bacterial counts from cleared and non-cleared leaf tissues by plating and incubating the samples at 25 °C for 3–7 days.
5. Perform partial 16S rRNA gene sequencing, according to Mahlangu et al. [12] to identify the isolates.

3.3.5 Microscopic Observations

1. Use an epifluorescence microscope (Axiophot; Zeiss) coupled with an AxioCam MRc5 camera to capture images.
2. Employ a confocal laser scanning microscope (LSM 510; Zeiss) to determine the spatial location of the bacteria inside the leaf tissues.
3. Visualize fluorescein using an argon laser at an excitation wavelength of 488 nm and visualize autofluorescence with a He-Ne laser (543 nm excitation wavelength).
4. Use superimposed section images to localize bacterial cells labeled with the bacteria-specific probe, scanning a series of 12 confocal sections with 1 μm increments.
5. Analyze the observed bacteria as microcolonies per mm^2 of the leaf tissue.

3.4 Fluorescence Confocal Microscopy

To study the spatial arrangement of endosymbionts within the host tissues, confocal microscopy can be used for 3D imaging [9].

3.4.1 DAPI Staining

1. Fix the cells in 3% paraformaldehyde in calcofluor white (CFW) at 4 °C for 1 h.
2. Wash the cells in cold 50 mM 4-(2-hydroxyethyl)-1-piperazineethanesulfonic acid (HEPES) buffer (pH 7.5) by centrifuging at 500×*g* for 30 s.
3. Pipette the cells onto microscope slides and air-dry them.
4. Incubate the samples with DAPI (4′,6-diamidino-2-phenylindole) staining solution (1 μg/mL) in 50 mM HEPES for 4 min.
5. Wash the slides twice with 50 mM HEPES buffer.

3.4.2 Photomicroscope and Image Analysis

1. Use a Zeiss AxioImager M1 microscope fitted with an LSM510 Meta confocal scan head and a Plan Apo 100/1.4 oil lens for confocal microscopy.
2. Excite DAPI with a 30-mW diode 405-nm laser, Cy3 with a 1.2-mW HeNe 543-nm laser, and Cy5 with a 5-mW HeNe 633-nm laser.
3. Visualize fluorescence signals using appropriate bandpass filters.
4. Obtain differential interference contrast (DIC) images with a 5-mW HeNe 633-nm laser.
5. Collect images using the Laser Scanning Microscope LSM510 software (version 4.0 SP2) and process them with Zeiss LSM Image Browser software (Version 4.0.0.12).
6. Use Zeiss LSM510 software to determine the colocalization of signals from EUB338 Cy3 and Cy5 in cells treated with these probes.

7. Calculate weighted colocalization coefficients, where a value of 0 indicates no fluorescence in the Cy3 channel colocalized with fluorescence from the Cy5 channel and a value of 1 indicates 100% colocalization of fluorescence between the Cy3 and Cy5 channels.

3.5 Transmission Electron Microscopy (TEM)

Transmission electron microscope allows the observation of the endosymbiont ultrastructure [13–15].

1. Fix dissected tissues in 2.5% glutaraldehyde in 0.1 M cacodylate buffer (pH 7.3) for 1 h.
2. Embed the tissues in a 2% agarose gel and fix them again in 2.5% glutaraldehyde in 0.1 M cacodylate buffer overnight.
3. Wash the tissues in 0.1 M cacodylate buffer for 20 min and repeat the wash three times.
4. Post-fix the tissues in 2% osmium tetroxide for 2 h, followed by washing and staining in 2% uranyl acetate for 90 min.
5. Dehydrate the tissues by serial immersion in ethanol (30%, 50%, 70%, 95%, and 100%) and transfer them to propylene oxide.
6. Embed the tissues in Epon resin and cut ultrathin sections (70 nm) using a diamond knife on a Leica Ultra cut UCT microtome.
7. Mount the sections on Pioloform-coated copper grids and stain them with saturated uranyl acetate followed by lead citrate.
8. View the sections using a Zeiss CEM 902 A transmission electron microscope at 80 kV.

3.6 Scanning Electron Microscopy (SEM)

3.6.1 Protocol A (Preparation of Bacteria for SEM) [12–15]

1. Centrifuge the bacterial broth.
2. Wash the pellet thrice with phosphate-buffered saline (PBS).
3. Add 0.25% glutaraldehyde (in sodium phosphate, pH 7.2) and incubate at room temperature for 30 min.
4. Incubate the samples overnight.
5. Wash the pellet thrice with phosphate buffer.
6. Collect the pellets by centrifugation.
7. Dehydrate the samples with ethanol at different concentrations (30%, 50%, 70%, 80%, 90%, and 100%) and incubate for 10 min at each step.
8. Incubate the samples in 100% ethanol for 1 h.
9. Prepare an SEM stub by applying adhesive tape and adding a bacterial sample to the tape.

3.6.2 Protocol B (Alternate Preparation Method for SEM)

1. Drop bacteria grown in broth onto polycarbonate (nucleopore) filter membranes (0.45 μm or 0.2 μm) and allow it to air-dry.
2. Fix the sample in 2.5% glutaraldehyde in PBS for 45 min to 1 h.
3. Rinse the samples in PBS for 15 min.
4. Re-fix the samples in 1% osmium tetroxide (OsO4) in PBS buffer for 1 h.
5. Dehydrate the samples through a series of ethanol rinses (30%, 50%, 70%, 80%, and 100%) for 10 min each.
6. Leave the samples in absolute ethanol for 15 min.
7. Perform critical point drying.
8. Attach the dried samples to an SEM stub.
9. Sputter coat the samples with gold (*see* **Note 9**).

3.6.3 The SEM Analysis

1. Grow bacterial strains in 5 mL of Luria broth overnight at 30 °C with shaking at 150 rpm.
2. Centrifuge the cultures at 1100×*g* for 10 min and discard the supernatant.
3. Briefly rinse the bacterial pellet with distilled water and fix it with a solution of 1% formaldehyde and 2% glutaraldehyde (1: 1, v/v) for 1 h at room temperature (25 °C) according to Schadler et al. [13].
4. Centrifuge the samples at 1100×*g* for 10 min and remove the supernatant.
5. Wash the pellet with 1000 μL of sterile distilled water.
6. Dehydrate the bacterial cells with increasing concentrations of ethanol (30%, 50%, 70%, 90%, 95%, and 100%) for 10 min each.
7. Store the dehydrated samples at 4 °C overnight.
8. Mount the dehydrated samples on SEM stubs and coat them with gold using an EmScope SC 500.
9. View the samples using a TESCAN VEGA SEM (VEGA 3 LHM, AVG9731276ZA) connected to a monitor.

4 Notes

1. Ensure that all collection tools are sterilized and handle plant samples gently to avoid damage. Collect samples during cooler parts of the day, keep them cool to maintain viability, and place plant material in sterile polyethylene bags or plastic containers.
2. Wash plant parts thoroughly with running water to remove contaminants. Carefully monitor the duration of surface

sterilization with sodium hypochlorite to avoid damaging the plant tissues and use control plates to verify sterilization effectiveness.

3. Maintain sterile techniques throughout the isolation process and clearly label all samples and agar plates for proper tracking of bacterial isolates.
4. Avoid overheating while fixing the smear, as it may distort the cell morphology.
5. Perform the staining step carefully, overstaining can lead to difficulty in decolorization.
6. Monitor the decolorization process closely; over-decolorizing can cause false negatives in Gram-positive cells, while under-decolorizing may leave Gram-negative cells stained purple.
7. Ensure to add oil immersion on the slide for accurate observations under the microscope at 100× objective lens, as it enhances clarity, especially for small or thin cells.
8. Select healthy leaves free of damage or stress to avoid any interference with FISH results.
9. Proper coating with gold is crucial for obtaining a high-resolution surface image, as this enhances the sample's conductivity.

Acknowledgement

This work was supported by the ZAR-NRF CSUR grant no. CSUR240404212233.

References

1. De Almeida CV, Andreote FD, Yara R, Tanaka FAO, Azevedo JL, de Almeida M (2009) Bacteriosomes in axenic plants: endophytes as stable endosymbionts. World J Microbiol Biotechnol 25:1757–1764
2. Patle PN, Navnage NP, Ramteke PR (2018) Endophytes in plant system: roles in growth promotion, mechanism and their potentiality in achieving agriculture sustainability. Int J Chem Stud 6:270–274
3. Jasim B, Joseph AA, John CJ, Mathew J, Radhakrishnan EK (2014) Isolation and characterization of plant growth promoting endophytic bacteria from the rhizome of Zingiber officinale. 3 Biotech 4:197–204
4. Alhaddad FA, Bitaar ZM, Abu-Dieyeh MH (2024) Diversity, characterization, and potential applications of bacterial endophytes isolated from the halophyte Limonium axillare. J Plant Growth Regul 43:2179–2196
5. Chauhan BP, Singh NK (2024) Morphological and biochemical characterization of endophytic bacteria from periwinkle [*Catharanthus roseus* (L.) G. Don]. Pharma Innov J 13:28–37
6. Kuechler SM, Renz P, Dettner K, Kehl S (2012) Diversity of symbiotic organs and bacterial endosymbionts of lygaeid bugs of the families blissidae and lygaeidae (Hemiptera: Heteroptera: Lygaeoidea). Appl Environ Microbiol 78:2648–2659
7. Cappuccino JG, Sherman N (1992) Microbiology: laboratory manual, 4th edn. Benjamin-Cummings Publishing, San Francisco
8. Lahlali R, Hijri M (2010) Screening, identification and evaluation of potential biocontrol fungal endophytes against Rhizoctonia solani

AG3 on potato plants. FEMS Microbiol Lett 311:152–159

9. Amann RI, Krumholz L, Stahl DA (1990) Fluorescent-oligonucleotide probing of whole cells for determinative, phylogenetic, and environmental studies in microbiology. J Bacteriol 172:762–770

10. Lo Piccolo S, Ferraro V, Alfonzo A, Settanni L, Ercolini D, Burruano S, Moschetti G (2010) Presence of endophytic bacteria in Vitis vinifera leaves as detected by fluorescence in situ hybridization. Ann Microbiol 60:161–167

11. Raina HS, Singh A, Popli S, Pandey N, Rajagopal R (2015) Infection of bacterial endosymbionts in insects: a comparative study of two techniques viz PCR and FISH for detection and localization of symbionts in whitefly, *Bemisia tabaci*. PLoS One 10(8):e0136159

12. Mahlangu SG, Serepa-Dlamini MH (2018) First report of bacterial endophytes from the leaves of *Pellaea calomelanos* in South Africa. S A J Sci 114(9/10). https://doi.org/10.17159/sajs.2018/4235

13. Golding CG, Lamboo LL, Beniac DR, Booth TF (2016) Scanning electron microscopy in microbiology and diagnosis of infectious diseases. Sci Rep 6:26516

14. Schadler R, Choi JH, Owens DK (2008) Application of electron microscopy to study the bacterial ultrastructure. Microsc Res Tech 71:437–450

15. Golding CG, Lamboo LL, Beniac DR, Booth TF (2016) The scanning electron microscope in microbiology and diagnosis of infectious disease. Sci Rep 6:26516

Chapter 13

PCR-Based Approaches to Study Endosymbionts

Kamlesh K. Yadav and Pratibha Yadav

Abstract

Endosymbionts play a major role in microbial ecology and host–microbe interactions, which has drawn a lot of attention to their molecular characterization. Numerous polymerase chain reaction (PCR)-based techniques are used to study endosymbiont genetic diversity and population dynamics. A useful technique for identifying genetic differences is PCR–restriction fragment length polymorphism (RFLP), which amplifies specific DNA fragments and then digests them with restriction enzymes. By examining terminal segments of bacterial 16S rRNA genes, the PCR–Terminal Restriction Fragment Length Polymorphism (T-RFLP) enhances community profiling. For microbial population research, PCR–Amplified Fragment Length Polymorphism (AFLP) is helpful since it provides thorough genomic fingerprints without requiring prior knowledge of the genome. PCR–Denaturing Gradient Gel Electrophoresis (DGGE) facilitates the detection of single-nucleotide polymorphisms (SNPs) by distinguishing DNA sequences according to their varying mobility under denaturing conditions. The structure and dynamics of microbial populations can be learned by the PCR–Temperature Gradient Gel Electrophoresis (TGGE) technique, which separates DNA molecules according to their melting patterns over a temperature gradient. By effectively investigating endosymbiont diversity, genetic variability, and evolutionary relationships, these molecular approaches contribute to gaining an understanding of microbial ecosystems.

Key words Endosymbiotic, PCR, Molecular techniques, RFLP, Fingerprinting, Restriction enzyme

1 Introduction

PCR-based methods are used in molecular analysis of endosymbionts which are microorganisms that exist in a symbiotic relationship with host organism. Different PCR methods enable specific information on the composition, roles, and phylogenetic placements of these symbionts. A very effective method of studying genetic variability among endosymbiont populations is called PCR–Restriction Fragment Length Polymorphism (RFLP) in which selective DNA segments are "amplified" by polymerase chain reaction and then "digested" using restriction enzymes to yield unique fragment profiles. In RFLP, the size of restriction fragments of a particular type of restriction enzyme varies between

N. Amaresan and Sapna Chandwani (eds.), *Endosymbionts*, Springer Protocols Handbooks,
https://doi.org/10.1007/978-1-0716-4771-4_13,

individuals and species to be used for the identification and investigation of genetic polymorphism as well as inter- and intraspecific differentiation. These patterns, or polymorphisms, can be used to distinguish between different strains or between different species of endosymbionts [1].

PCR–Terminal Restriction Fragment Length Polymorphism (T-RFLP) is a similar technique that targets the terminal fragments of the amplified DNA making it convenient for comparative analysis of microbial community structure across different host and or environmental conditions. This technique followed the PCR amplification of a selected fragment of bacterial 16S rRNA genes directly from the total community DNA, where one of the two PCR primers contained a fluorescent dye at its 5′ end. The PCR product was then subjected to restriction digestion, and the detected terminal restriction fragment was fluorescence-labeled and accurately quantified with an automated DNA sequencer. It is fast and efficient and may be used to assess the diversity and composition of microbial community [2, 3].

Another method is called PCR–Amplified Fragment Length Polymorphism (AFLP) which is a differential and more sophisticated technique based on the selective amplification of restriction fragments to create and compare unique genomic fingerprints and to evaluate variation within and among microbial populations, including endosymbionts. This technique generates distinctive banding patterns for genetic profiling by combining the selective PCR amplification of restriction fragments with selective primers and gel electrophoresis [4]. Most importantly, AFLP does not require any prior knowledge of the genome sequence of the species under investigation, which is a strength of the technique [5].

Two of the PCR-based methods which do this are the PCR–Denaturing Gradient Gel Electrophoresis (DGGE) and PCR–Temperature Gradient Gel Electrophoresis (TGGE). PCR–DGGE is a molecular approach to studying microbial communities such as endosymbionts according to the sequence differences of genes that have been amplified by PCR. This method resolves DNA fragments with different sequences depending on the separation of the DNA fragments based on the melting action of the chemical gradient composed of urea and formamide within the gel. During electrophoresis, DNA fragments move through the gel and, at some point, are subjected to gradually increasing concentrations of denaturing agents, and hence, they are only partially denatured if the former were fully double-stranded before they entered the gel. This leads to differing mobility, which means that fragments of the same size but with different sequences can be separated. It has practical applications for the detection of single-nucleotide polymorphisms (SNP), without the need to sequence the genome, making it a particularly useful method for analyzing microbial communities especially the 16S rRNA genes [6].

PCR–Temperature Gradient Gel Electrophoresis (TGGE) process resolves DNA fragments according to melting profile in a temperature gradient gel allowing the study of community structure and dynamics. The particularity of this technique is based on melting where the double helix with two strands of DNA is separated based on its sensitivity to temperature. Non-sequence artifacts of the DNA separation technique reveal that a molecule's size and composition determine the mobility of DNA in the gel and allow the separation of DNA of similar molecular weights but with different sequences. TGGE can be conducted in two modes depending on the orientation of the electric field relative to the temperature gradient: The common or parallel mode where a number of samples can be analyzed at once is again different from the other mode that allows detailed analysis of the sample, which is referred to as perpendicular mode.

2 Materials

2.1 PCR–RFLP

- Tris 1 M (pH 0.8): Tris base, 12.11 g; dH_2O, 100 mL, adjust pH 8.0 with NaOH EDTA.
- Tris–EDTA (10 mM Tris–HCl, 1 mM EDTA, pH 8): Tris (pH 8.0), 10 mM; EDTA, 1 mM.
- Lysozyme (100 mg/mL).
- Proteinase (20 mg/mL).
- Sodium dodecyl sulfate (SDS, 10%): 10 g SDS; dH_2O, adjust final volume 100 mL.
- 5 M sodium chloride (NaCl): NaCl, 29.25 g, dissolve in dH_2O, adjust final volume 100 mL.
- CTAB–NaCl: CTAB, 10 g; 0.7 M NaCl, 100 mL, adjust final 100 mL CTAB–NaCl.
- Chloroform–isoamyl alcohol (24:1): 48 mL of chloroform; 2 mL of isoamyl alcohol for 50 mL.
- Isopropanol (ice cold).
- Agarose gel (1%): agarose, 1 g; Et–Br, 5 μL; add dH_2O to 100 mL.
- Ethidium bromide (Et–Br): Et–Br, 10 mg; water, 1 mL.
- Ethanol chilled (70%): 70 mL ethanol; add 30 mL water.
- TAE buffer (10×): Tris base, 48.4 g; glacial acetic acid, 11.42 mL; EDTA (0.5 M), 20 mL.
- Water bath.
- Centrifuge.
- Picodrop.

- UV transilluminator.
- Gel Documentation System.
- Thermal cyclers (PCR).
- PCR amplification kit (reagents as supplied):

  10× Taq buffer; 2.5 mM $MgCl_2$; 10 pM forward primer; 10 pM reverse primer; 10 mM dNTPs; 5 U Taq polymerase; nuclease-free water.
- Primer forward/reverse.
- Restriction enzyme kit (reagents as supplied):

  10× buffer R; restriction enzyme; nuclease-free water.
- Restriction enzyme: MvaI; EcoRI; SphI; MspI.
- Gel electrophoresis unit.

2.2 PCR–T-RFLP

- Tris 1 M (pH 0.8): Tris base, 12.11 g; dH_2O, 100 mL, adjust pH 8.0 with NaOH EDTA.
- Tris–EDTA (10 mM Tris–HCl, 1 mM EDTA, pH 8): Tris (pH 8.0), 10 mM; EDTA, 1 mM.
- Lysozyme (100 mg/mL).
- Proteinase (20 mg/mL).
- Sodium dodecyl sulfate (SDS, 10%): 10 g SDS; dH_2O, adjust final volume 100 mL.
- 5 M sodium chloride (NaCl): NaCl, 29.25 g, dissolve in dH_2O, adjust final volume 100 mL.
- CTAB–NaCl: CTAB, 10 g; 0.7 M NaCl, 100 mL, adjust final 100 mL CTAB–NaCl.
- Chloroform–isoamyl alcohol (24:1): 48 mL of chloroform; 2 mL of isoamyl alcohol for 50 mL.
- Isopropanol (ice cold).
- Agarose gel (1%): agarose, 1 g; Et–Br, 5 μL; add dH_2O to 100 mL.
- Ethidium bromide (Et–Br): Et–Br, 10 mg; water, 1 mL.
- Ethanol chilled (70%): 70 mL ethanol; add 30 mL water.
- Water bath.
- Centrifuge.
- Picodrop.
- UV transilluminator.
- Gel Documentation System.
- Thermal cyclers (PCR).
- PCR amplification kit (reagents as supplied):

10× Taq buffer; 2.5 mM $MgCl_2$; 10 pM forward primer; 10 pM reverse primer; 10 mM dNTPs; 5 U Taq polymerase; nuclease-free water.

- Fluorescent-labeled forward/reverse primer.
- Restriction enzyme kit (reagents as supplied):

  10× buffer R; restriction enzyme; nuclease-free water.
- TAE buffer (10×): Tris base, 48.4 g; glacial acetic acid, 11.42 mL; EDTA (0.5 M), 20 mL.
- Gel electrophoresis unit.
- Capillary electrophoresis unit.
- GENESCAN software.

2.3 PCR–AFLP

- Tris 1 M (pH 0.8): Tris base, 12.11 g; dH_2O, 100 mL, adjust pH 8.0 with NaOH EDTA.
- Tris–EDTA (10 mM Tris–HCl, 1 mM EDTA, pH 8): Tris (pH 8.0), 10 mM; EDTA, 1 mM.
- Lysozyme (100 mg/mL).
- Proteinase (20 mg/mL).
- Sodium dodecyl sulfate (SDS, 10%): 10 g SDS; dH_2O, adjust final volume 100 mL.
- 5 M sodium chloride (NaCl): NaCl, 29.25 g, dissolve in dH_2O, adjust final volume 100 mL.
- CTAB–NaCl: CTAB, 10 g; 0.7 M NaCl, 100 mL, adjust final 100 mL CTAB–NaCl.
- Chloroform–isoamyl alcohol (24:1): 48 mL of chloroform; 2 mL of isoamyl alcohol for 50 mL.
- Isopropanol (ice cold).
- Agarose gel (1%): agarose, 1 g; Et–Br, 5 μL; add dH_2O to 100 mL.
- Ethidium bromide (Et–Br): Et–Br, 10 mg; water, 1 mL.
- Ethanol chilled (70%): 70 mL ethanol; add 30 mL water.
- Water bath.
- Centrifuge.
- Picodrop.
- UV transilluminator.
- Gel Documentation System.
- Thermal cyclers (PCR).
- MseI adaptor.
- EcoRI adaptor.
- Reagents for ligation of adapters (reagents as supplied):

EcoRI adapter; MseI adapter; 10× T4 DNA ligase buffer; T4 DNA ligase (3 U/μL); ddH_2O (double-distilled water).

- PCR amplification reaction kit (reagents as supplied):

  EcoRI primer; MseI primer; dNTPs (5 mM); 10× PCR buffer; $MgCl_2$; ddH_2O (double-distilled water); Taq polymerase (5 U/μL).

- Sequi-Gen GT sequencing cell.
- Silver staining with a commercial kit (reagents as supplied).
- Polyacrylamide gel (6%): 6.0 g acrylamide; 0.25 g bisacrylamide; 42.2 g urea; 2.5 mL 20× TBE; 50 mL ddH_2O; mix and bring volume to 100 mL.
- TAE buffer (10×): Tris base, 48.4 g; glacial acetic acid, 11.42 mL; EDTA (0.5 M), 20 mL.
- 10× TBE (Tris–borate–EDTA) buffer: Tris base 108 g, boric acid 55 g, 40 mL 0.5 M EDTA, adjust volume 1000 mL with distilled water.
- 10 bp DNA ladder.
- 100 bp DNA ladder.

2.4 PCR–DGGE

- Tris 1 M (pH 0.8): Tris base, 12.11 g; dH_2O, 100 mL, adjust pH 8.0 with NaOH EDTA.
- Tris–EDTA (10 mM Tris–HCl, 1 mM EDTA, pH 8): Tris (pH 8.0), 10 mM; EDTA, 1 mM.
- Lysozyme (100 mg/mL).
- Proteinase (20 mg/mL).
- Sodium dodecyl sulfate (SDS, 10%): 10 g SDS; dH_2O, adjust final volume 100 mL.
- 5 M sodium chloride (NaCl): NaCl, 29.25 g, dissolve in dH_2O, adjust final volume 100 mL.
- CTAB–NaCl: CTAB, 10 g; 0.7 M NaCl, 100 mL, adjust final 100 mL CTAB–NaCl.
- Chloroform–isoamyl alcohol (24:1): 48 mL of chloroform; 2 mL of isoamyl alcohol for 50 mL.
- Isopropanol (ice cold).
- Agarose gel (1%): agarose, 1 g; Et–Br, 5 μL; add dH_2O to 100 mL.
- Ethidium bromide (Et–Br): Et–Br, 10 mg; water, 1 mL.
- Ethanol chilled (70%): 70 mL ethanol; add 30 mL water.
- Water bath.
- Centrifuge.
- Picodrop.

- UV transilluminator.
- Gel Documentation System.
- Thermal cyclers (PCR).
- PCR amplification kit (reagents as supplied):

  10× Taq buffer; 2.5 mM $MgCl_2$; 10 pM forward primer; 10 pM reverse primer; 10 mM dNTPs; 5 U Taq polymerase; nuclease-free water.
- Primer forward/reverse.
- Nested-PCR–DGGE reagents:
- Reaction buffer ($MgCl_2$-free) 10×; $MgCl_2$, 1.5 mM; primer forward (P338F with added 5′ GC-clamp), primer reverse (P518r); deoxynucleoside triphosphate, 200 μM each; bovine serum albumin, 400 ng μL^{-1}; Taq DNA polymerase, 1.25 U/ 50 μL; DNase- and RNase-free filter-sterilized water.
- TAE buffer (10×): Tris base, 48.4 g; glacial acetic acid, 11.42 mL; EDTA (0.5 M), 20 mL.
- Polyacrylamide gels (45–60%): 100% denaturant contains 7 M urea and 40% formamide.
- Ethanol (10%): 10 mL ethanol; add 90 mL water.
- Acetic acid (0.5%): 0.5 mL glacial acetic acid; add 95.5 mL water.
- Bionumerics software 2.0 (Applied Maths, Kortrijk, Belgium) for DGGE gel band pattern analysis.

2.5 PCR–TGGE

- Tris 1 M (pH 0.8): Tris base, 12.11 g; dH_2O, 100 mL, adjust pH 8.0 with NaOH EDTA.
- Tris–EDTA (10 mM Tris–HCl, 1 mM EDTA, pH 8): Tris (pH 8.0), 10 mM; EDTA, 1 mM.
- Lysozyme (100 mg/mL).
- Proteinase (20 mg/mL).
- Sodium dodecyl sulfate (SDS, 10%): 10 g SDS; dH_2O, adjust final volume 100 mL.
- 5 M sodium chloride (NaCl): NaCl, 29.25 g, dissolve in dH_2O, adjust final volume 100 mL.
- CTAB–NaCl: CTAB, 10 g; 0.7 M NaCl, 100 mL, adjust final 100 mL CTAB–NaCl.
- Chloroform–isoamyl alcohol (24:1): 48 mL of chloroform; 2 mL of isoamyl alcohol for 50 mL.
- Isopropanol (ice cold).
- Agarose gel (1%): agarose, 1 g; Et–Br, 5 μL; add dH_2O to 100 mL.

- Ethidium bromide (Et–Br): Et–Br, 10 mg; water, 1 mL.
- Ethanol chilled (70%): 70 mL ethanol; add 30 mL water.
- Water bath.
- Centrifuge.
- Picodrop.
- UV transilluminator.
- Gel Documentation System.
- Thermal cyclers (PCR).
- PCR amplification kit (reagents as supplied):

  10× Taq buffer; Tris–HCl (pH 8.3), 10 mM; KCl, 50 mM; $MgCl_2$, 3 mM; deoxynucleoside triphosphate, 0.2 mM each; Taq polymerase, 1.25 U, primers forward and reverse, 10 pmol.
- Primers for specific V6 to V8 regions of the 16S rRNA:

  U968-GC (59-CGC CCG GGG CGC GCC CCG GGC GGG GCG GGG GAA CGG GAA GAA CCT TAC-39) and L1401 (59-GCG TGT GTA CAA GAC CC-39).
- Casting mold for gel solution preparation for gel electrophoresis.
- Polyacrylamide gel stock (**30%**)—acrylamide:bisacrylamide (37.5:1):
  - 37 g of acrylamide monomer; add 1 g of bisacrylamide (cross-linker) and transfer to a 150-mL graduated cylinder containing ~40 mL of dH_2O and mix for about 30 min.
  - Heat up the solution to 70 °C for the first 10 min.
  - Bring the volume to 100 mL with dH_2O and filter through the Whatman filter paper.
  - Store at 4–10 °C, in a glass bottle away from light.
- 7 M urea:
  - Add 42.04 g of urea in approximately 60–70 mL of 1.25× Tris–acetate buffer (pH 8.0).
  - Stir until fully dissolved.
  - Adjust final volume 100 mL with 1.25× Tris–acetate buffer.
- Gel solution for TGGE (8% PAGE): concentration 8% w/v acrylamide:bisacrylamide 37.5:1, 7 M urea) in 1.25× Tris–acetate buffer (pH 8).
  - Slowly add 26.67 mL of 30% acrylamide (37.5:1) stock solution to the urea solution while stirring gently.
  - Add an additional 1.25× Tris–acetate buffer to bring the total volume to 100 mL.

- Ammonium persulfate (10% w/v): Add 1 g of ammonium persulfate to ddH_2O and bring the volume to 10 mL. Store at 4 °C.
- Tetra methylene diamine (TEMED): (Bio-Rad or Sigma). Store at 4 °C.
- TAE buffer (10×): Tris base, 48.4 g; glacial acetic acid, 11.42 mL; EDTA (0.5 M), 20 mL.
- 10× TBE (Tris–borate–EDTA) buffer: Tris base 108 g, boric acid 55 g, 40 mL 0.5 M EDTA, adjust volume 1000 mL with distilled water.
- SYBR green.
- Ethidium bromide solution.

3 Methods

3.1 PCR–RFLP

3.1.1 DNA Extraction

1. The isolation of genomic DNA is done by using the protocol suggested by Sambrook et al. [7] with slight modifications [7].
2. Centrifuge the overnight-grown fresh bacteria culture in nutrient broth at 5000×*g* for 5 min.
3. Resuspend pellet in Tris–EDTA (10 mM Tris–HCl, 1 mM EDTA, pH 8).
4. Repeat heat shock method (freezing and thawing) for lysis of bacterial cells.
5. Add 20 μL lysozyme (100 mg/mL); incubate at 37 °C for 30 min, followed by the addition of 3 μL proteinase K (20 mg/mL); and incubate at 42 °C for 60 min.
6. Subsequently, add 25 μL SDS (10%) followed by 80 μL of 5 M NaCl and 80 μL of CTAB–NaCl.
7. Mix properly and incubate at 65 °C for 10 min, then allow to cool to room temperature.
8. Add an equal volume of chloroform–isoamyl alcohol (24:1) and mix gently and centrifuge at 9000×*g* for 10 min at 4 °C.
9. Collect the upper aqueous phase in a fresh tube and add an equal volume of chloroform–isoamyl alcohol (24:1) again.
10. Centrifuge the mixture at 9000×*g* for 10 min at 4 °C.
11. Collect the aqueous phase again in a new tube and add 0.6 volume of ice-cold isopropanol to it.
12. Mix gently and incubate at 30–60 min, and then, centrifuge at 12,000×*g* for 10 min at 4 °C.
13. Discard the supernatant and wash the pellet twice with 70% chilled ethanol.
14. Finally, allow the DNA pellet to air-dry and suspend in 50 μL Tris–EDTA buffer [7–9].

3.1.2 Quantitative and Qualitative Analysis of DNA

1. Quantify the DNA using Picodrop (model: *PICO* 100 Cambridge, UK) at 260 nm.
2. The ratio of absorbance at 260 nm to 280 nm is used for determining the purity of the DNA.
3. Additionally, agarose gel electrophoresis techniques (0.8% agarose gel) are also used to evaluate the quality of the DNA.
4. Weigh the agarose 0.8 g and dissolve it in the TAE buffer until the agarose looks transparent.
5. Allow to cool at room temperature and add the ethidium bromide (10 mg/mL) to it.
6. Load the dissolved agarose gel into the gel electrophoresis unit and put the comb for making the well.
7. Allow the gel to solidify at normal temperature.
8. Fill the electrophoresis unit with TAE buffer and load the DNA along with the loading buffer.
9. Allow the current to migrate the DNA and visualize the DNA fragments under UV light in the UV transilluminator.
10. Store the DNA images on a computer by the Gel Documentation System (Alpha Innotech, USA).
11. Keep the extracted DNA at −20 °C for further use [10, 11].

3.1.3 PCR Amplification

1. Amplify the small subunit of the 16S rRNA gene of approximately 1.5 kb DNA segment using primer sets 16S1 (5′--GAGTTTGATCCTGGCTCA-3′) and 16S2 (5′-CGGCTACCTTGTTACGACTT-3′) using a G-STORM (Gene Technology Thermal Cyclers) [12].
2. Prepare the reaction mixture in 25 μL volume using a PCR amplification kit (Qiagen, India) as a final concentration (Table 1) [8].

The program for PCR reaction was set as follows:

- Initial denaturation: 94 °C for 2.5 min.
- 35 cycles of
  - Denaturation: 94 °C for 30 seconds.
  - Annealing: 55 °C for 30 seconds (depending on primer Tm).
  - Extension: 72 °C for 1 min.
- Final extension: 72 °C for 7 min.

3.1.4 Quality and Quantity of Amplified PCR Product

1. Prepare 1.2% agarose gel to check the quality and quantity of the amplified PCR product.
2. Run the gel electrophoresis and visualize the band staining with ethidium bromide (final concentration 0.5 μg/mL).

Table 1
The reaction mixture for PCR

Reagent	Volume (25 μL)
Taq buffer (10×)	2.5
MgCl2 (2.5 mM)	1.5
Primer forward (10 pM)	0.5
Primer forward (10 pM)	0.5
dNTP (10 mM)	0.5
Taq polymerase (5 U)	0.25
Nuclease-free water	18.25
Template (DNA)	1.0

Table 2
Restriction enzymes used for cutting 16S rRNA gene of all isolates with their cutting sites

S. No.	Restriction enzyme	Cutting site
1	MvaI	5′ CC↓WGG 3′ 3′ GGW↑CC 5′
2	EcoRI	5′ G↓AATTC 3′ 3′ CTTAA↑G 5′
3	SphI	5′ GCATG↓C 3′ 3′ C↑GTACG 5′
4	MspI	5′ C↓CGG 3′ 3′ GGC↑C 5′

3.1.5 Restriction Enzyme Digestion

1. Use different restriction enzymes having different restriction sites for the analysis of restriction fragment length polymorphism (RFLP) patterns of the 16S rRNA gene of bacterial isolates (Table 2) [11].
2. Purify the 16S rRNA PCR amplicons using the PCR purification kit as per the manufacturer's instructions (Qiagen, India).
3. Prepare the digestion reaction as mentioned in Table 3 to digest through restriction enzymes according to the manufacturer's instructions (Thermo Fisher Scientific, UK).
4. Mix all reagents gently and spin down for a few seconds.
5. Incubate at 37 °C for 1–16 h.

Table 3
Recommended protocol for digestion

Reagent volume	(25 μL)
Nuclease-free water	16 μL
10× buffer R	2 μL
DNA (0.5–1 μg/μL)	1 μL
MvaI	0.5–2 μL

3.1.6 Gel Electrophoresis

1. Prepare a 2–3% agarose gel stained with a DNA intercalating agent such as ethidium bromide.
2. Load the digested PCR products in respective wells of the gel along with a DNA ladder.
3. Run the gel at an appropriate voltage that is usually between 80 and 120 V until the dyes reach a certain distance.
4. Expose it to UV light to visualize the DNA fragments and take a picture.

3.1.7 Pattern Analysis

1. Compare the banding pattern to analyze the polymorphism among the samples.
2. The patterns of differences describe the different strains or species of endosymbionts [9, 11].

3.2 PCR–T-RFLP

3.2.1 DNA Extraction

Isolate the DNA as per the instructions described in Subheading 3.1.1 [3, 8, 13].

3.2.2 Quantitative and Qualitative Analysis of DNA

Analyze quantitative and qualitative DNA as per instructions described in Subheading 3.1.2.

3.2.3 PCR Amplification

1. Follow all the instructions described in Subheading 3.1.3 for amplification of 16S rRNA, except primer pairs.
2. Use the primer pairs that have been labeled with fluorescent tags and follow the instructions of manufacturers for the PCR amplification of DNA.

The program for PCR reaction was set as follows:

- Initial denaturation: 94 °C for 2.5 min.
- 35 cycles of
  - Denaturation: 94 °C for 30 s.
  - Annealing: 55 °C for 30 s (depending on primer Tm).
  - Extension: 72 °C for 1 min.
- Final extension: 72 °C for 7 min.

3.2.4 Quality and Quantity of Amplified PCR Product

1. Prepare the 1.2% agarose gel and run gel electrophoresis at an appropriate voltage that is usually between 80 and 120 V until the dyes reach a certain distance.
2. Visualize the DNA band through staining with ethidium bromide (final concentration 0.5 μg/mL).
3. Expose it to UV light to check the DNA quality and quantity of the amplified PCR product.

3.2.5 Restriction Enzyme Digestion

1. Purify the 16S rRNA PCR amplicons using the PCR purification kit as per the manufacturer's instructions (Qiagen, India).
2. Prepare the digestion reaction to digest through restriction enzymes according to the manufacturer's instructions (Thermo Fisher Scientific, UK).
3. Take 200–300 ng of purified PCR product (assuming a 50% loss during purification) and 10–20 U of restriction enzyme for each digestion.
4. The incubation period for restriction enzymes varies from 4 to 12 h at 37 °C to ensure complete digestion (refer to Restriction Enzyme Digestion Subheading 3.1.5).
5. Heat the reaction mixture at 65 °C for 20–25 min to inactivate the restriction enzymes.
6. Restriction enzyme digestion generates varying lengths of products with respect to the sequence diversity of PCR products.

3.2.6 Capillary Electrophoresis

1. Use capillary electrophoresis to differentiate terminal restriction fragments by the size of the fragments.
2. Separate the digested products by capillary electrophoresis on an automated sequencer.
3. Run the electrophoretic as indicated by the manufacturer of the instrument.
4. Fragments can be detected since the primers used in the procedure are labeled with a fluorescent material [2].

3.2.7 Data Analysis T-RFLP Profiles

1. Evaluate fragment lengths and peak intensities with the help of a special program [14].
2. Obtain a T-RFLP profile that would give an indication of its structure.

   Two ways of variation may be among the profile generated by T-RFLP:

   - The first possibility is that the number and size (base pairs) of T-RFs present in a profile may differ.
   - Secondly, the variation may be found in the height (and consequently the area) of any particular peak (*see* **Note 1**).

3. The second numerical result produced by the analysis program is a table that lists each peak's size in base pairs (calculated by reference to the internal standard) and the height of each peak (relative to the amount of fluorescence detected).
4. Define a minimum threshold of fluorescence first to exclude background noise when analyzing any particular profile.
5. Set the minimum peak height normally at 50–100 fluorescent units. Use the Local Southern size calling algorithm method on the GENESCAN software to determine the T-RF size following a comparison with the internal standard [2].

3.3 PCR–AFLP

3.3.1 DNA Extraction

Isolate the DNA as per the instructions described in Subheading 3.1.1 [3, 8, 13].

3.3.2 Quantitative and Qualitative Analysis of DNA

Analyze quantitative and qualitative DNA as per instructions described in Subheading 3.1.2.

3.3.3 Preparing the AFLP Template

1. Process the genomic DNA with a pair of restriction enzymes, often MseI and EcoRI. MseI is characterized by a four-base pair cutting site, and EcoRI is characterized by six base pairs, as mentioned in the above table. MseI and EcoRI generate DNA pieces with 5′ overhangs (5′TA3′ and 5′AATT3′, respectively) that are distinct from each other and are non-complementary.
2. Incubate the genomic DNA with these enzymes for a period long enough to permit complete digestion of the DNA into restriction fragments. The incubation period for restriction enzymes varies up to 16 h at 37 °C to ensure complete digestion (refer to Restriction Enzyme Digestion Subheading 3.1.5).

3.3.4 Ligation Reaction with Restriction Fragments and Adaptors

1. Use an MseI adaptor and an EcoRI adaptor for ligation, each of which is a short, double-stranded DNA molecule with a 5′ overhang that is complementary to the 5′ overhang generated by MseI (5′TA3′) and EcoRI (5′AATT3′), respectively, as per the manufacturer's instructions (Promega Corporation).
2. Prepare the ligation reaction as per Table 4.
3. Add 10 μL of the ligation mix to 50 μL of MseI/EcoRI digested genomic DNA.
4. Vortex briefly and spin down if necessary.
5. Incubate at room temperature for 3 h.
6. Add the MseI and EcoRI restriction enzymes in the ligation reaction, to stop relegation between EcoRI and MseI genomic fragments and will not recognize the ligated adaptor-genomic DNA fragments.
7. Agitate every hour [15].

Table 4
Ligation of adapters

Reagent	Volume (10 μL)
EcoRI adapter	1.0 mL
MseI adapter	1.0 mL
T4 DNA ligase 10× buffer	1.0 mL
T4 DNA ligase (3 U/mL, Promega)	0.3 mL
ddH_2O	6.7 mL

Table 5
Preamplification of 2 μL of restriction ligation product combined with the following

Reagent	Volume (20 μL)
EcoRI + A oligo primer (50 ng/mL)	0.5 μL
MseI + C oligo primer (50 ng/mL)	0.5 μL
dNTPs (5 mM)	2.0 μL
10× PCR buffer (comes with Taq)	2.0 μL
$MgCl_2$ (comes with Taq)	1.2 μL
ddH_2O	11.7 μL
Template DNA from restriction/ligation	2.0 μL
Taq polymerase (5 U/mL)	0.1 μL

3.3.5 Preamplification Reactions

1. Prepare the reaction mixture for preamplification according to the table below (Table 5), using 2 μL of restriction ligation product.
2. Adjust the PCR thermal cycle conditions as mentioned below.

   The program for PCR reaction was set as follows:

- Initial denaturation: 94 °C for 2 min.
- 35 cycles of
  - Denaturation: 94 °C for 1 min.
  - Annealing: 56 °C for 1 min.
  - Extension: 72 °C for 1 min.
- Final extension: 72 °C for 5 min [15].

3.3.6 Selective PCR Amplification

1. Add the preamplification PCR product to 100 μL of sterile water.
2. Add 1 μL of dilute (6:1) preamplification PCR product to 19 μL of selective amplification mixture (25 ng *EcoR*I primer, 30 ng *Mse*I primer, 0.4 mM dNTPs, 1× PCR buffer, 0.4 U Taq polymerase, 1.5 mM $MgCl_2$).
3. Set the PCR thermal cycle condition as per the given below.

The program for PCR reaction was set as follows:

Initial denaturation at 94 °C for 2 min.

Followed by 12 cycles with annealing temperatures decreasing by 0.7 °C per cycle (94 °C for 30 s, 65 °C for 30 s, 72 °C for 1 min).

Then 23 cycles.

(94 °C for 30 s, 56 °C for 30 s, 72 °C for 1 min); final extension at 72 °C for 2 min, ending with a soak at 4 °C

3.3.7 Gel Electrophoresis

1. Use capillary electrophoresis or a denaturing polyacrylamide gel electrophoresis to separate the AFLP products.
2. Add 8 mL of formamide loading buffer to the PCR product for denaturing polyacrylamide gel electrophoresis.
3. Run the electrophoresis unit for 3 h at 75 W on a 6% polyacrylamide gel cast on a 38 × 50 cm Sequi-Gen GT sequencing cell (Bio-Rad Laboratories Inc., Hercules, CA, USA).
4. Use a 10 bp ladder (Gibco BRL #10821-015) as a molecular weight marker in two lanes.
5. Conduct silver staining with a commercial kit (Promega #Q4132) according to instructions, except that both the fix/-stop and developing solutions were partially frozen at −20 °C to a slurry.
6. Observe the banding patterns visually, then transfer the gels on chromatography paper and allow them to dry.

3.3.8 Analysis of Data

1. Examine banding patterns to determine polymorphisms and genetic variation within endosymbiont populations.
2. Apply specialist software for statistical analysis and fragment size scoring [14].

3.4 PCR-DGGE

3.4.1 DNA Extraction

Isolate the DNA as per the instructions described in Subheading 3.1.1 [3, 8, 13].

3.4.2 Quantitative and Qualitative Analysis of DNA

Analyze quantitative and qualitative DNA as per instructions described in Subheading 3.1.2.

3.4.3 First-Round PCR Amplification

Follow all the instructions described in Subheading 3.1.3 for amplification of 16S rRNA.

3.4.4 Second-Round Nested PCR–DGGE

1. Prepare the reaction mixture for the amplification in the second-round nested-PCR amplification.
2. Add 1 μL of amplified product from the first-round PCR amplification to 49 μL of PCR mixture described below in Table 6 [16].
3. Set the PCR thermal cycle condition as per the given below.
4. After each PCR amplification round, the size of the PCR product was verified on a 1% agarose gel.

The program for nested PCR reaction was set as follows:

- Initial denaturation: 95 °C for 10 min.
- 30 cycles of
  - Denaturation: 94 °C for 1 min.
  - Annealing: 53 °C for 1 min.
  - Extension: 72 °C for 2 min.
- Final extension: 72 °C for 12 min [17].

Table 6
The final concentrations of the different components in the master mix (50 μL)

Reagents	Concentration	Stock conc.	Volume to add
Primer (P338F[a], P518r)	0.2 μM of each	10 μM	1.0 μL
Deoxynucleoside triphosphate each	200 μM	10 mM each	1.0 μL
$MgCl_2$	1.5 mM	25 mM	3.0 μL
Reaction buffer ($MgCl_2$-free)	10×	10×	5.0 μL
Taq DNA polymerase	1.25 U/50 μL (Promega, Madison, WI, USA)	5 U/μL	0.25 μL
Bovine serum albumin	400 ng μL^{-1}	20,000 ng/μL	1.0 μL
DNase- and RNase-free filter-sterilized water			37.75 μL

See **Note 2**

3.4.5 DGGE Run Process

1. Use the Bio-Rad D Gene System (Bio-Rad, Hercules, CA, USA) to carry out DGGE following the protocol of Muyzer et al. [18].
2. Place the second-round nested-PCR amplicon onto 8% (w/v) polyacrylamide gels in 1× TAE (20 mM Tris, 10 mM acetate, 0.5 mM EDTA pH 7.4).
3. Prepare the polyacrylamide gels using a denaturing gradient that ranges from 45% to 60% (for bacteria and ammonium oxidizers) or from 50% to 65% (for actinomycetes, methanotrophs, acidobacterium, and α-Proteobacteria) (where 100% denaturant contains 7 M urea and 40% formamide).
4. Run electrophoresis for up to 16 h at 60 °C and at 37 V for bacteria and α-Proteobacteria, or 40 V for actinomycetes, methanotrophs, and acidobacterium.
5. After electrophoresis, immerse the gels in fixing buffer (10% ethanol and 0.5% acetic acid) for 5 min, followed by 30 min of ethidium bromide solution immersion, wash in Milli-Q water, and capture the images in a Bio-Rad Gel Doc station [16, 17, 19–21].

3.4.6 Analysis of DGGE Patterns

1. Use Bionumerics software 2.0 (Applied Maths, Kortrijk, Belgium) to process the DGGE gel band pattern analysis. The similarity calculation generates a distance matrix and is based on the Pearson (product–moment) correlation coefficient [22].
2. Use Bionumerics 2.0 to carry out clustering analysis, multidimensional scaling (MDS), and principal component analysis (PCA).
3. Use Ward's clustering algorithm to compute the dendrograms of each DGGE gel, as well as the combination of all gels [23].
4. Employ MDS and PCA to simplify the data from the complex DGGE patterns of one sample into a single point in three-dimensional space.
5. Instead of analyzing the original data set, MDS uses a similarity coefficient (Pearson's correlation) to analyze each DGGE's distance matrix.
6. Because PCA involves direct data analysis, it differs from MDS. However, the only way to process fingerprint types for PCA is to first create a band-matching table.
7. Each band is categorized into classes of common bands, and each pattern can have two states for a given band class: present and missing (binary matrix).

3.5 PCR–TGGE to Study Endosymbionts

3.5.1 DNA Extraction

1. Extract DNA from host tissues containing the endosymbionts using a commercial extraction kit [3].

3.5.2 Quantitative and Qualitative Analysis of DNA

1. Assess DNA quality and quantity suitable for PCR amplification as per instructions described in Subheading 3.1.2.

3.5.3 PCR Amplification

1. Use the primer for the amplification of the specific V6 to V8 regions of the 16S rRNA, using primers U968-GC (59-CGC CCG GGG CGC GCC CCG GGC GGG GCG GGG GAA CGG GAA GAA CCT TAC-39) and L1401 (59-GCG TGT GTA CAA GAC CC-39) [24].
2. A 30- to 40-bp GC clamp is included in one of the PCR primers to ensure that the region under analysis is positioned within the lower melting domain and that the DNA mostly remains double-stranded.
3. Allow for the separation of the PCR products generated by the U968-GC primers, which include a GC clamp, using TGGE [18].
4. Prepare the 50 μL PCR reaction mixture according to the instructions and composition listed in Table 7.
5. Set the PCR thermal cycling conditions according to the instructions given below.
6. Run conventional electrophoresis on a 2% agarose gel containing ethidium bromide to confirm the sizes and quantities of the amplicons after PCR amplification using aliquots of 10 μL.

Table 7
Concentrations of the components in the PCR reaction mix (50 μL) [25]

Reagents	Concentration
Tris–HCl (pH 8.3)	10 mM
KCl	50 mM
$MgCl_2$	3 mM
Deoxynucleoside triphosphate (each)	0.2 mM
Taq polymerase (Roche)	1.25 U
Primers	10 pmol
DNA template	5 mL

The program for PCR reaction was set as follows:

- Initial denaturation: 95 °C for 5 min.
- 30 cycles of
  - Denaturation: 94 °C for 1 min.
  - Annealing: 42 °C for 1 min.
  - Extension: 68 °C for 1 min.
- Final extension: 68 °C for 7 min [25].

3.5.4 TGGE Procedure

1. Follow the various steps described by Viglasky (2013) and Manzano et al. (2000) to carry out the TGGE process [25, 26].
2. Assemble the casting mold according to the instructions provided.
3. Prepare the gel solution for gel electrophoresis (0.8-mm polyacrylamide gel) at the desired polyacrylamide concentration of 8% w/v (acrylamide:bisacrylamide 37.5:1, 7 M urea) in 1.25 M Tris–acetate buffer (pH 8).
4. Add ammonium persulfate and TEMED to achieve final concentrations of 0.05% each, mix well, and pour immediately into the gel casting mold.
5. Fill the gel tank with the 0.5× TBE gel running buffer and load a DNA sample into the well.
6. Allow the sample to migrate into the gel for 15 min at approximately 5 V/cm, without a temperature gradient.
7. Interrupt the run for 30–40 min and establish a temperature gradient.
8. Apply a temperature ramp of 0.5 °C/h from 67 to 69 °C during the electrophoresis.
9. Run the gel at 5–10 V/cm for 2–7 h with a constant voltage.
10. Stop the run and proceed with one of the staining protocols described below.
11. Stain the gel in 1.25 M Tris–acetate buffer containing 1× SYBR Green (Sigma, Milan, Italy) or ethidium bromide solution, followed by immersion in Milli-Q water, and capture images with a DS34 camera system (Polaroid, St. Louis, MO) [25].

4 Notes

1. This variation can have a major influence on estimates of the biodiversity represented in the numerically rarer members of the community. This variance can be seen when comparing the peak heights in TRFLP profiles from different microbial

compositions. The height of each peak can provide a measure of the relative proportion of each component of a population, although biases caused by preferential annealing of the primer of templates mean that the absolute values should be treated with caution [27].

2. Add 5′ GC-clamp to primer P338F for DGGE analysis [18].

References

1. Chaudhary R, Maurya GK (2019) Restriction fragment length polymorphism. In: Encyclopedia of animal cognition and behavior. Springer Nature, Cham. https://doi.org/10.1007/978-3-319-47829-6_175-1
2. Tiquia SM (2010) Using terminal restriction fragment length polymorphism (T-RFLP) analysis to assess microbial community structure in compost systems. Methods Mol Biol 599:89–102. https://doi.org/10.1007/978-1-60761-439-5_6
3. Liu WT, Marsh TL, Cheng H, Forney LJ (1997) Characterization of microbial diversity by determining terminal restriction fragment length polymorphisms of genes encoding 16S rRNA. Appl Environ Microbiol 63(11): 4516–4522
4. Chial H (2008) DNA fingerprinting using amplified fragment length polymorphisms (AFLP): no genome sequence required. Nat Educ 1(1):176
5. Paun O, Schönswetter P (2012) Amplified fragment length polymorphism: an invaluable fingerprinting technique for genomic, transcriptomic, and epigenetic studies. Methods Mol Biol 862:75–87. https://doi.org/10.1007/978-1-61779-609-8_7
6. Strathdee F, Free A (2013) Denaturing gradient gel electrophoresis (DGGE). Methods Mol Biol 1054:145–157. https://doi.org/10.1007/978-1-62703-565-1_9
7. Sambrook J, Russell DW (2001) Molecular cloning: a laboratory manual, 3rd edn. Cold Spring Harbor Laboratory Press, New York
8. Yadav KK, Bora A, Datta S, Chandel K, Gogoi HK, Prasad GB, Veer V (2015a) Molecular characterization of midgut microbiota of Aedes albopictus and Aedes aegypti from Arunachal Pradesh, India. Parasit Vectors 18(8): 641. https://doi.org/10.1186/s13071-015-1252-0. PMID: 26684012; PMCID: PMC4683861
9. Yadav KK, Chandel K, Bora A, Veer V (2015) Isolation and characterization of Dietzia maris from midgut of Aedes albopictus: a suitable candidate for paratransgenesis. Int J Mos Res 2(4):7–12
10. Yadav KK, Datta S, Naglot A, Bora A, Hmuaka V, Bhagyawant S et al (2016a) Diversity of cultivable midgut microbiota at different stages of the Asian tiger mosquito, Aedes albopictus from Tezpur, India. PLoS ONE 11(12): e0167409. https://doi.org/10.1371/journal.pone.0167409
11. Yadav KK, Naglot A, Chandel K, Bhagyawant S, Bora A, Gogoi HK, Vijay V, Raju PS (2016b) Molecular characterization of midgut bacteria of Aedes albopictus exhibiting swarming motility property. Afr J Microbiol Res 10(25):949–959. https://doi.org/10.5897/AJMR2016.8068
12. Alam SI, Dixit A, Reddy GS, Dube S, Palit M, Shivaji S et al (2006) Clostridium schirmacherense sp. nov., an obligately anaerobic, proteolytic, psychrophilic bacterium isolated from lake sediment of Schirmacher Oasis, Antarctica. Int J Syst Evol Microbiol 56:715–720
13. Gupta N (2019) DNA extraction and polymerase chain reaction. J Cytol 36(2):116–117. https://doi.org/10.4103/JOC.JOC_110_18. PMID: 30992648; PMCID: PMC6425773
14. Hulce D, Liu C (J) (2006) GeneMarker® software for terminal-restriction fragment length polymorphism (T-RFLP) data analysis. Soft Genetics LLC. SoftGenetics LLC App Note. www.softgenetics.com
15. Hazen SP, Leroy P, Ward R (2002) AFLP in *Triticum aestivum* L.: patterns of genetic diversity and genome distribution. Euphytica 125(1):89–102
16. Boon N, Windt WD, Verstraete W, Top EM (2002) Evaluation of nested PCR–DGGE (denaturing gradient gel electrophoresis) with group-specific 16S rRNA primers for the analysis of bacterial communities from different wastewater treatment plants. FEMS Microbiol Ecol 39(2):101–112. https://doi.org/10.1111/j.1574-6941.2002.tb00911.x

17. Ovreås L, Forney L, Daae FL, Torsvik V (1997) Distribution of bacterioplankton in meromictic Lake Saelenvannet, as determined by denaturing gradient gel electrophoresis of PCR-amplified gene fragments coding for 16S rRNA. Appl Environ Microbiol 63(9): 3367–3373. https://doi.org/10.1128/aem.63.9.3367-3373.1997
18. Muyzer G, de Waal EC, Uitterlinden AG (1993) Profiling of complex microbial populations by denaturing gradient gel electrophoresis analysis of polymerase chain reaction-amplified genes coding for 16S rRNA. Appl Environ Microbiol 59(3):695–700. https://doi.org/10.1128/aem.59.3.695-700.1993
19. Muyzer G, Small K (1998) In microbial ecology, the use of DGGE and TGGE. Thus the contributions of: Antonie van Leeuwenhoek 73:127–141
20. Dar SA, Kuenen JG, Muyzer G (2005) Nested PCR-denaturing gradient gel electrophoresis approach to determine the diversity of sulfate-reducing bacteria in complex microbial communities. Appl Environ Microbiol 71(5): 2325–2330. https://doi.org/10.1128/AEM.71.5.2325-2330.2005
21. Ercolini D (2004) PCR-DGGE fingerprinting: new approaches towards identification of microorganisms present in foods. J Microbiol Methods 56:297–314
22. Pearson K (1926) On the coefficient of radical likeliness. Biometrika 18:105–117
23. Ward JH (1963) Hierarchical grouping to optimize an objective function. J Am Stat Assoc 58: 236–244
24. Nübel U, Engelen B, Felske A, Snaidr J, Wieshuber A, Amann RI, Ludwig W, Backhaus H (1996) Sequence heterogeneities of genes encoding 16S rRNAs in Paenibacillus polymyxa detected by temperature gradient gel electrophoresis. J Bacteriol 178(19): 5636–5643. https://doi.org/10.1128/jb.178.19.5636-5643.1996
25. Manzano M, Cocolin L, Cantoni C, Comi G (2000) Temperature gradient gel electrophoresis of the amplified product of a small 16S rRNA gene fragment for the identification of listeria species isolated from food. J Food Prot 63(5):659–661. https://doi.org/10.4315/0362-028X-63.5.659
26. Viglasky V (2013) Polyacrylamide temperature gradient gel electrophoresis. In: Makovets S (ed) DNA electrophoresis: methods and protocols, vol 1054. Springer, pp 159–172. https://doi.org/10.1007/978-1-62703-565-1_10
27. Suzuki MT, Giovannoni SJ (1996) Bias caused by template annealing in the amplification mixtures of 16S rRNA genes by PCR. Appl Environ Microbiol 62:625–630

Chapter 14

Non-PCR-Based Approaches to Study Endosymbionts

T. S. Archana, Devendra Kumar, Shivam Singh, and Gagan Kumar

Abstract

Understanding the role and dynamics of endosymbionts within host organisms is crucial in various fields including ecology, evolutionary biology, and biotechnology. Traditional PCR-based methods have significantly contributed to this understanding, but they often pose limitations such as the inability to distinguish between live and dead cells or the requirement for prior knowledge of target sequences. Non-PCR-based approaches offer innovative solutions to overcome these challenges and provide deeper insights into the ecology and physiology of endosymbionts. Fluorescence In Situ Hybridization (FISH) and Catalyzed Reporter Deposition–Fluorescence In Situ Hybridization (CARD–FISH), flow cytometry and cell sorting, and DNA microarray technology are some of the important non-PCR-based techniques. FISH and CARD–FISH allow for the visualization and localization of specific endosymbionts within host tissues or environmental samples, enabling researchers to examine spatial relationships and quantify cellular abundances with high resolution. Flow cytometry, coupled with cell sorting, facilitates the isolation and purification of endosymbiont-containing cells based on their physical and chemical properties, thus enabling downstream genomic or proteomic analyses. DNA microarrays provide a high-throughput method to analyze the genetic diversity and functional capabilities of endosymbiont communities by simultaneously detecting multiple genetic targets.

Key words Non-PCR, Endosymbionts, Ecology

1 Introduction

Endosymbionts, the microorganisms that live within plant tissues, play a crucial role in plant health and disease resistance, making their study vital in plant pathology. These microorganisms enhance nutrient acquisition, with certain endosymbionts like mycorrhizal fungi facilitating the uptake of essential nutrients such as phosphorus and nitrogen, which boosts plant growth and resilience [1, 2]. Some endosymbionts produce growth-promoting substances that increase plant development and stress tolerance. Additionally, endosymbionts can induce systemic resistance in plants, making them less susceptible to infections by various pathogens. Understanding the interactions between endosymbionts and their

N. Amaresan and Sapna Chandwani (eds.), *Endosymbionts*, Springer Protocols Handbooks,
https://doi.org/10.1007/978-1-0716-4771-4_14,

plant hosts is essential for developing sustainable agricultural practices and bio-based solutions for plant disease management, ultimately improving crop yields and overall plant health. While PCR (polymerase chain reaction) has been a fundamental tool in microbial studies, it has significant limitations in endosymbiont research [3]. Issues with sensitivity and specificity can lead to false positives or negatives due to contamination or nonspecific primer binding. Quantification challenges arise because qPCR measures total DNA, including both live and dead cells, potentially misrepresenting the active microbial population. The complexity of microbial communities further complicates PCR-based methods, as they can struggle to differentiate closely related species or strains. Additionally, primer selection can introduce bias, favoring the amplification of certain sequences over others, resulting in incomplete or skewed representations of the microbial community. Non-PCR-based approaches have emerged as valuable alternatives for studying endosymbionts, offering solutions to the limitations of PCR. These methods provide comprehensive insights into the roles and interactions of endosymbionts in plant pathology [4]. Fluorescence In Situ Hybridization (FISH) and Catalyzed Reporter Deposition–FISH (CARD–FISH) are prominent techniques. FISH uses fluorescently labeled probes that bind to specific nucleic acid sequences within cells, allowing for the visualization and identification of endosymbionts within plant tissues while preserving their spatial context. CARD–FISH enhances FISH by amplifying the signal, making it more sensitive and suitable for detecting low-abundance endosymbionts [3].

Flow cytometry and cell sorting are other powerful techniques. Flow cytometry analyzes the physical and chemical characteristics of cells in a mixture, and when combined with fluorescent markers, it can differentiate and quantify various endosymbionts within a sample [5]. Cell-sorting techniques enable the isolation of specific cell populations based on their fluorescence properties, facilitating a detailed study of individual endosymbionts and their functions [6]. DNA microarrays, which consist of thousands of DNA probes on a solid surface, can simultaneously measure the expression levels of numerous genes, providing a comprehensive view of the genetic activity within endosymbiont populations [7]. This approach is especially useful for studying gene expression profiles under different environmental conditions, shedding light on the roles of endosymbionts in plant health and disease resistance. These non-PCR-based approaches are robust tools for studying endosymbionts in plant pathology, offering deeper insights into their diversity, functions, and interactions within the plant host. By utilizing these methods, researchers can develop more effective strategies for managing plant diseases and enhancing crop resilience, contributing to more sustainable agricultural practices.

2 Materials

2.1 Fish

1. Endosymbiont-containing tissue or cell samples
2. Phosphate-buffered saline (PBS) (*see* **Note 1**)
3. Paraformaldehyde or formaldehyde (*see* **Note 2**)
4. Ethanol or methanol
5. Fluorescently labeled oligonucleotide probes
6. Hybridization buffer (with formamide, salts, blocking agents)
7. Microscope slides and coverslips
8. Humidified chamber
9. Water bath or incubator
10. Stringency washing buffers (various concentrations of SSC) (*see* **Note 3**)
11. Anti-fade mounting medium (e.g., VECTASHIELD)
12. Fluorescence microscope with appropriate filters
13. Pipettes and sterile pipette tips
14. Parafilm or other sealing materials
15. Gloves, lab coat, and PPE

2.2 Card–Fish

1. Endosymbiont-containing tissue or cell samples
2. Phosphate-buffered saline (PBS)
3. Paraformaldehyde or formaldehyde
4. Ethanol or methanol
5. Horseradish peroxidase (HRP)-labeled oligonucleotide probes
6. Hybridization buffer (with formamide, salts, blocking agents)
7. Tyramide signal amplification (TSA) kit
8. Hydrogen peroxide (H_2O_2)
9. Microscope slides and coverslips
10. Humidified chamber
11. Water bath or incubator
12. Stringency washing buffers (various concentrations of SSC)
13. Anti-fade mounting medium (e.g., VECTASHIELD)
14. Fluorescence microscope with appropriate filters
15. Pipettes and sterile pipette tips
16. Parafilm or other sealing materials
17. Gloves, lab coat, and PPE (*see* **Note 4**)

2.3 Catalyzed Reporter Deposition–Fluorescence In Situ Hybridization

1. Endosymbiont-containing tissue or cell samples
2. Phosphate-buffered saline (PBS)
3. Paraformaldehyde or formaldehyde
4. Ethanol or methanol
5. Horseradish peroxidase (HRP)-labeled oligonucleotide probes
6. Hybridization buffer (containing formamide, salts, and blocking agents)
7. Tyramide signal amplification (TSA) reagents
8. Hydrogen peroxide (H_2O_2)
9. Sodium chloride (NaCl)
10. Tris–HCl buffer
11. Microscope slides and coverslips
12. Humidified chamber
13. Water bath or incubator
14. Stringency washing buffers (various concentrations of SSC)
15. Anti-fade mounting medium (e.g., VECTASHIELD)
16. Fluorescence microscope with appropriate filters
17. Pipettes and sterile pipette tips
18. Parafilm or other sealing materials
19. Gloves, lab coat, and PPE

2.4 Flow Cytometry

1. Endosymbiont-containing cell suspension
2. Phosphate-buffered saline (PBS)
3. Fluorescent dyes or fluorochrome-conjugated antibodies
4. Fixatives (e.g., paraformaldehyde)
5. Permeabilization agents (e.g., saponin or Triton X-100)
6. Blocking reagents (e.g., bovine serum albumin, BSA)
7. Flow cytometry tubes or microtiter plates
8. Flow cytometer with appropriate laser and filters
9. Sterile pipettes and pipette tips
10. Centrifuge and centrifuge tubes
11. Cell strainer or filters
12. Gloves, lab coat, and PPE

2.5 Cell-Sorting Approach to Study Endosymbionts

1. Endosymbiont-containing cell suspension
2. Phosphate-buffered saline (PBS)
3. Fluorescent dyes or fluorochrome-conjugated antibodies specific to endosymbionts
4. Fixatives (e.g., paraformaldehyde)

5. Permeabilization agents (e.g., saponin or Triton X-100)
6. Blocking reagents (e.g., bovine serum albumin, BSA)
7. Cell sorting buffer (PBS with BSA or FBS)
8. Flow cytometer with cell sorting capabilities (fluorescence-activated cell sorting, FACS)
9. Sterile flow cytometry tubes
10. Sterile pipettes and pipette tips
11. Centrifuge and centrifuge tubes
12. Collection tubes (e.g., microcentrifuge tubes or multi-well plates)
13. Cell strainer or filters
14. Gloves, lab coat, and PPE

3 Methods

3.1 Fish

1. Collect and fix samples with 4% paraformaldehyde or ethanol for 2–4 h at 4 °C
2. Permeabilize samples using lysozyme or proteinase K, depending on the target organism
3. Prepare the hybridization buffer with a fluorescently labeled probe, formamide, salts, and detergents
4. Incubate samples with the buffer at 42–46 °C for 2–3 h in a humid chamber
5. Wash samples with a buffer at 46 °C for 15–20 min to remove unbound probes
6. Counterstain with DAPI to visualize host structures (optional)
7. Mount samples on a slide with an anti-fade mounting medium
8. Visualize under a fluorescence microscope using appropriate filters
9. Analyze fluorescence intensity and distribution to determine endosymbiont presence and localization [8]

3.2 Card–Fish

1. *Sample Fixation:* Collect and fix samples containing endosymbionts using 4% paraformaldehyde at 4 °C for 2–4 h. Rinse samples with PBS and store in ethanol–PBS solution (1,1) at –20 °C until use.
2. *Permeabilization:* Permeabilize samples with lysozyme (10 mg/mL in 0.05 M EDTA and 0.1 M Tris–HCl, pH 7.5) (*see* **Note 5**) for gram-positive bacteria or proteinase K (15 μg/mL in 0.1 M Tris–HCl, pH 8.0) for 1 h at 37 °C to enhance probe penetration.

3. *Inactivation of Endogenous Peroxidases:* Treat samples with 0.15% hydrogen peroxide in methanol for 30 min at room temperature to inactivate endogenous peroxidases that could interfere with the amplification step. Rinse thoroughly with PBS.
4. *Hybridization:* Prepare the hybridization buffer containing horseradish peroxidase (HRP)-labeled oligonucleotide probe, 0.9 M NaCl, 20% formamide, 0.01% SDS, and 20 mM Tris–HCl (pH 7.4). Apply buffer to samples and allow to incubate at 35–46 °C for 2–3 h in a humid chamber (*see* **Note 6**).
5. *Washing:* Rinse samples in pre-warmed washing buffer containing 0.9 M NaCl, 5 mM EDTA, and 20 mM Tris–HCl (pH 7.4) for 10–20 min at 46 °C to remove unbound probes.
6. *Signal Amplification:* Incubate samples with amplification buffer containing fluorescently labeled tyramide (e.g., Alexa Fluor dyes) and hydrogen peroxide for 10–30 min at room temperature to enhance signal. Stop the reaction by rinsing with PBS.
7. *Counterstaining (optional):* Stain samples with DAPI to visualize host nuclei or other structures, and rinse thoroughly.
8. *Mounting:* Mount samples on slides using an anti-fade mounting medium to prevent photobleaching.
9. *Microscopy:* Examine samples under a fluorescence microscope using appropriate filters to detect HRP-probe-conjugated fluorescence. Capture images for analysis.
10. *Analysis:* Quantify the fluorescence intensity and distribution to assess the presence, localization, and abundance of endosymbionts in the host tissue [9].

3.3 Flow Cytometry

1. *Sample Preparation:* Collect samples containing endosymbionts and dissociate them into single cells using mechanical disruption or enzymatic digestion (e.g., cellulase or pectinase for plant tissues).
2. *Fixation:* Fix cells with 4% paraformaldehyde or 70% ethanol at 4 °C for 30 min to preserve cellular structures. Wash cells with PBS to remove excess fixative (*see* **Note 7**).
3. *Permeabilization:* Permeabilize cells using 0.1% Triton X-100 or saponin in PBS for 10–15 min at room temperature to allow staining probes to enter.
4. *Staining:* Incubate cells with fluorescently labeled probes or antibodies specific to endosymbionts (e.g., FISH probes or antibodies against endosymbiont-specific proteins) in the dark for 30–60 min at 4 °C. Wash with PBS to remove unbound probes or antibodies.

5. *Counterstaining (optional):* Stain cells with DAPI or another DNA-binding dye to visualize the host cell nuclei. Rinse thoroughly with PBS.
6. *Filtering:* Pass the stained cell suspension through a 40-µm cell strainer to remove clumps and aggregates that could interfere with flow cytometry analysis.
7. *Flow Cytometry:* Run the prepared samples through a flow cytometer. Use appropriate laser settings and detection filters to detect the fluorescence signals corresponding to the endosymbiont markers and host cell markers [9].
8. *Data Acquisition:* Acquire data from at least 10,000 events per sample to ensure statistical significance. Record the fluorescence intensity, forward scatter (FSC), and side scatter (SSC) to differentiate endosymbionts from host cells.
9. *Data Analysis:* Analyze data using flow cytometry software to identify and quantify endosymbionts based on specific fluorescence signals and scatter properties. Use gating strategies to distinguish endosymbionts from debris and host cells.
10. *Validation (optional):* Validate results by comparing them with microscopic observations or molecular methods (e.g., qPCR) to confirm the presence and abundance of endosymbionts.

3.4 Cell-Sorting Approach

1. *Sample Collection and Preparation:* Collect samples containing endosymbionts and disaggregate them into single-cell suspensions using mechanical disruption or enzymatic digestion (e.g., cellulase for plant tissues or collagenase for animal tissues).
2. *Fixation (optional):* If needed, fix cells with 4% paraformaldehyde or 70% ethanol for 30 min at 4 °C to preserve cellular integrity. Wash cells with PBS to remove excess fixative.
3. *Permeabilization (if required):* Permeabilize cells using 0.1% Triton X-100 or saponin for 10–15 min at room temperature to enable the entry of staining probes or antibodies.
4. *Staining:* Incubate cells with fluorescently labeled probes or antibodies specific to endosymbionts (e.g., FISH probes or antibodies against endosymbiont-specific proteins) for 30–60 min at 4 °C in the dark. Wash cells with PBS to remove unbound probes or antibodies [10].
5. *Counterstaining (optional):* Stain cells with a DNA-binding dye such as DAPI to visualize host cell nuclei or other structural components. Rinse thoroughly with PBS.
6. *Filtering:* Pass the stained cell suspension through a 40-µm cell strainer to remove aggregates and debris, ensuring a uniform single-cell suspension.

7. *Flow Cytometry and Sorting:* Use a cell sorter with a flow cytometer to identify and sort cells based on specific fluorescence signals and scatter properties. Set gates to differentiate endosymbiont-containing cells from host cells and debris (*see* **Note 8**).
8. *Sorting:* Sort target cells based on the defined fluorescence criteria into collection tubes or multi-well plates. Ensure aseptic conditions if live cell recovery is required.
9. *Collection and Verification:* Collect sorted cells and verify their purity and viability. Reanalyze a fraction of the sorted cells to confirm the accuracy of sorting. Use microscopy or PCR-based methods for further validation.
10. *Downstream Applications:* Use sorted endosymbiont-containing cells for downstream analyses, such as genomic sequencing, transcriptomics, proteomics, or functional assays to study their role and interactions within the host [11].

3.5 DNA Microarray

1. *Sample Collection:* Collect host tissues or cells containing endosymbionts. Ensure the samples are kept on ice or at −80 °C to prevent degradation.
2. *Nucleic Acid Extraction:* Extract total DNA or RNA from the samples using a suitable extraction kit or method (e.g., phenol–chloroform extraction or column-based kits). Include steps to ensure the separation of host and endosymbiont DNA or RNA, if possible. Quantify and assess the purity of the extracted nucleic acids using spectrophotometry (e.g., NanoDrop) or fluorometry (e.g., Qubit).
3. *cDNA Synthesis (for RNA samples):* Convert extracted RNA to complementary DNA (cDNA) using reverse transcription with random hexamers or oligo(dT) primers. Purify the synthesized cDNA using a purification kit.
4. *Labeling of Nucleic Acids:* Label the extracted DNA or synthesized cDNA with fluorescent dyes (e.g., Cy3 or Cy5) using a random priming or direct labeling method. Incorporate dyes to ensure each sample has a unique fluorescent label. Purify the labeled nucleic acids to remove any unincorporated dyes.
5. *Microarray Preparation:* Use a prefabricated microarray slide containing probes specific to endosymbiont genes or design a custom microarray with oligonucleotide probes representing known or predicted endosymbiont genes. Ensure the microarray slides are pre-treated and ready for hybridization [12] (*see* **Note 9**).
6. *Hybridization:* Mix labeled DNA or cDNA samples with a hybridization buffer containing formamide, salts, and blocking agents. Apply the mixture onto the microarray slide and cover it

with a coverslip. Incubate the slide in a hybridization chamber at 42–65 °C for 12–24 h to allow specific binding of labeled nucleic acids to complementary probes.

7. *Washing:* Wash the microarray slides in a series of washing buffers with decreasing salt concentrations (e.g., SSC buffer) to remove nonspecifically bound nucleic acids. The final wash should be in low salt concentration to reduce background noise. Dry the slides by centrifugation or air-drying in a dust-free environment.
8. *Scanning:* Scan the microarray slides using a microarray scanner equipped with lasers that excite the specific fluorescent dyes (e.g., Cy3 and Cy5). Adjust the scanner settings to optimize signal detection while minimizing background noise. Save the scanned images as high-resolution files for analysis [13].
9. *Data Analysis:* Use microarray analysis software to analyze the scanned images. Perform background correction and normalization (e.g., using quantile normalization), and quantify fluorescence intensity for each probe. Identify differentially expressed genes by comparing the intensity values between samples.
10. *Validation (optional):* Validate microarray results using complementary methods such as quantitative PCR (qPCR) to confirm the expression levels of selected genes.
11. *Interpretation:* Analyze the data to understand the functional roles of endosymbionts, their interactions with the host, and their contributions to host physiology or adaptation. Interpret results in the context of biological pathways, gene functions, or microbial communities.

4 Notes

1. While preparing the Tris buffer, add water at the bottom of the container to dissolve the Tris easily, allowing the magnetic stir bar to work immediately. If using a glass beaker, warm the water to about 37 °C to dissolve Tris faster, but ensure the solution is cooled to room temperature before adjusting the pH.
2. Follow safety measures for handling and disposing of all reagents, particularly hazardous chemicals.
3. Prepare all hybridization and washing buffers in advance using the above guidelines. Ensure proper pH adjustment to enhance probe binding and prevent pH fluctuations that may affect the hybridization process.

4. Use appropriate personal protective equipment (PPE) at all times, including masks, gloves, lab coats, and eye protection.
5. Adjust the pH of buffers using concentrated HCl (12 N) initially to reach close to the desired pH. Use lower concentrations (e.g., 6 N, 1 N) for fine adjustments to prevent a sudden drop in pH below the target.
6. Prepare all staining buffers and solutions with precise pH control using appropriate HCl concentrations as noted earlier. Ensure potentially hazardous reagents are handled with care, following fume hood protocols and PPE guidelines.
7. Handle all hazardous materials, particularly in well-ventilated areas or fume hoods.
8. Dispose of all hazardous waste following the manufacturer's guidelines and institutional safety protocols to ensure a safe working environment.
9. When handling large batches of reagents, consider storing aliquots as described to maintain consistency and reproducibility in experimental conditions.

References

1. Karabulut F, Parray JA, Shafi N, Ikram M (2024) Endophytes: role in maintaining plant health under stress conditions. In: Plant endophytes and secondary metabolites. Academic Press, London, pp 105–132
2. Xia Y, Liu J, Chen C, Mo X, Tan Q, He Y, Zhou G (2022) The multifunctions and future prospects of endophytes and their metabolites in plant disease management. Microorganisms 10:1072. https://doi.org/10.3390/microorganisms10051072
3. Chandwani S, Gohil K, Dwivedi MK, Amaresan N (2023) In situ techniques to study non-cultivable microbial symbionts. In: Microbial symbionts. Academic Press, London, pp 765–785
4. Badr A, Fouad W (2023) Unlocking the richness of microalgae biodiversity for potential applications. In: Microalgae-current and potential applications. IntechOpen, London
5. McKinnon KM (2018) Flow cytometry: an overview. Curr Protoc Immunol 120:5–1. https://doi.org/10.1002/cpim.40
6. Orfao A, Ruiz-Argüelles A (1996) General concepts about cell sorting techniques. Clin Biochem 29:5–9. https://doi.org/10.1016/0009-9120(95)02079-7
7. Stoughton RB (2005) Applications of DNA microarrays in biology. Annu Rev Biochem 74:53–82. https://doi.org/10.1146/annurev.biochem.74.082803.133211
8. Rocha R, Almeida C, Azevedo NF (2018) Influence of the fixation/permeabilization step on peptide nucleic acid fluorescence in situ hybridization (PNA-FISH) for the detection of bacteria. PLoS One 13(5):e0196522
9. Pernthaler A, Pernthaler J, Amann R (2002) Fluorescence in situ hybridization and catalyzed reporter deposition for the identification of marine bacteria. Appl Environ Microbiol 68: 3094–3101. https://doi.org/10.1128/AEM.68.6.3094-3101.2002
10. Hattersley SM et al (2012) Microfluidic tissue dissociation for single-cell suspension preparation: chemical and mechanical methods for high viability. Lab Chip 12:2375–2383. https://doi.org/10.1039/c2lc40216b
11. Lombardo JA et al (2021) Integrated microfluidic tissue dissociation device for efficient single-cell recovery. Nat Commun 12:6809. https://doi.org/10.1038/s41467-021-27082-x
12. Schena M et al (1995) Quantitative monitoring of gene expression patterns with a complementary DNA microarray. Science 270:467–470. https://doi.org/10.1126/science.270.5235.467
13. Chuaqui RF et al (2002) Post-analysis follow-up and validation of microarray experiments. Nat Genet 32(Suppl):509–514. https://doi.org/10.1038/ng1033

Chapter 15

Single Cell–Based Technologies to Study Endosymbionts: Raman Microspectroscopy

Komal Chandarana and Natarajan Amaresan

Abstract

Raman microspectroscopy can be used to analyze live, single endosymbiotic cells with minimal sample preparation, preserving cellular integrity. It is a label-free technique that is ideal for studying endosymbionts, where traditional staining may alter or destroy cellular components. This present chapter describes the detailed preparation of the sample and Raman microscope setup along with short data processing information.

Key words Endosymbiotic cell, Label-free technique, Raman microspectroscopy, Sample preparation

1 Introduction

Investigating endosymbionts within host organisms has yielded significant insights into the core aspects of symbiosis, evolution, and cellular interactions. These internal symbionts play a crucial role in managing various physiological functions of the host, including the acquisition of nutrients, modulation of immune responses, and reactions to stress. Despite their importance, the complex interactions of endosymbionts within host cells remain largely unexplored due to the lack of high-resolution analytical methods [1]. Conventional bulk analysis techniques cannot capture the single-cell-level dynamics. Examining endosymbionts at the individual cell level is crucial to elucidate their specific functions and metabolic exchanges among cells in a population [1]. Raman microspectroscopy has emerged as a powerful, noninvasive imaging technique among cutting-edge technologies [2]. This allows researchers to investigate the metabolic and molecular characteristics of both hosts and their endosymbiotic partners without the need to culture the holobiont [3].

Raman microspectroscopy employs the principle of inelastic scattering of single-wavelength light to yield comprehensive

N. Amaresan and Sapna Chandwani (eds.), *Endosymbionts*, Springer Protocols Handbooks,
https://doi.org/10.1007/978-1-0716-4771-4_15,

insights into molecular vibrations [4]. This technique enables the examination of the biochemical makeup and metabolic conditions of individual cells and their associated endosymbionts [3]. In this chapter, the detailed sample preparation, instrumental setup, and Raman spectroscopy imaging have been described to facilitate the molecular characterization of endosymbionts from various hosts according to the method described by Yang et al. and Haessler et al. [3, 5].

2 Materials

2.1 Sample (Fresh or Fixed Tissue or Cell Samples Containing Endosymbiont)

2.2 Chemicals and Reagents

1. Phosphate-buffered saline.
2. 4% paraformaldehyde.
3. Silver or gold nanoparticles (optional).
4. Standard biomolecules (lipid, protein, nucleic acids) for reference spectra.

2.3 Glassware and Equipment

1. Microscopic slides (made up of quartz, calcium fluoride, or standard stainless steel MALDI plate).
2. High-precision coverslip.
3. Cooling centrifuge.
4. Confocal Raman microscope equipped with CCD camera.

2.4 Software

1. Software capable of multivariate analysis (MATLAB, Origin, ProjectFOUR (WITec, Oxford Instrument), ImageJ).

3 Methods

3.1 Sample Preparation

1. Collect the host organism harboring endosymbiont (insect, algae, sponge, protist, or other tissue) ensuring minimal contamination from external microbes by handling the specimen in a sterile environment.
2. Dissect the host organism if necessary and transfer it into sterile phosphate-buffered saline (PBS).
3. Homogenize the material gently using a glass homogenizer to release the endosymbionts from the host.

4. Centrifuge the suspension at 3000 rpm for 10 min at 4 for 10 min at 4 °C to remove large debris and collect the supernatant containing endosymbiont in the fresh centrifuge tube.
5. Wash the suspension twice in PBS at 5000 rpm for 10 min at 4 °C.

3.2 Sample Fixation

1. Discard the supernatant from the above step and fix the cells using 4% paraformaldehyde for 10 min (*see* **Note 1**).
2. Centrifuge the paraformaldehyde-fixed sample at 5000 rpm for 10 min.
3. Remove the supernatant and resuspend the cell suspension with PBS by gentle vortexing.
4. Repeat **steps 2** and **3** twice to remove residual fixative and keep cell suspension at 4 °C.

3.3 Preparations Prior to Sample Loading

1. Dilute the purified sample in PBS to achieve a concentration suitable for single-cell analysis (*see* **Note 2**).
2. Collect the 10 μL of the suspension on a clean, low-background glass slide made up of quartz or calcium fluoride or a standard stainless steel MALDI plate.
3. Remove the excess medium with the tissue paper (*see* **Note 3**).
4. Allow the suspension to air-dry under the laminar air flow hood.
5. For surface-enhanced Raman scattering (SERS), apply silver or gold nanoparticles to the sample to enhance signal intensity. Mix nanoparticles with the cell suspension before drying.

3.4 Instrument Setup

1. Select appropriate laser wavelengths that are generally used to analyze biological samples (e.g., 532 nm, 633 nm, 785 nm) in a confocal Raman microscope (*see* **Notes 4** and **5**).
2. Set the high-NA (numerical aperture) objective (60x or 100x oil immersion) to focus the laser on individual cells.
3. Ensure the high-sensitivity detector (CCD camera is installed for signal acquisition).
4. Select a grating optimized for the spectral range of interest (400–2000 cm^{-1}).
5. Calibrate the instrument using silicon (Raman-standard material) with a prominent peak at 520.7 cm^{-1}.
6. Focus the laser precisely on individual cells using brightfield or fluorescence imaging (*see* **Note 6**).

3.5 Raman Spectroscopic Imaging

1. Adjust acquisition time (e.g., 0.1–5 s) to balance signal intensity and photodamage risk (*see* **Note 5**).

2. Record Raman spectra for the cell. Ensure baseline correction and fluorescence subtraction.

3.6 Data Processing

The outlined steps for data processing are described for the software ProjectFOUR (WITec, Oxford Instrument) and MATLAB. However, it can be done using other software for spectral processing also.

1. Background subtraction: Elimination or minimization of background signals in acquired Raman data sets can be achieved using polynomial fits or WITec ProjectFOUR's "shape" function which operates based on a rolling circle background subtraction.
2. Cosmic ray removal: Eliminate Raman peaks originating from cosmic rays by applying appropriate settings (*see* **Note 7**).
3. Export/import: Use the Graph ASCII file format to export the data file from the acquisition software and import it into MATLAB.
4. Reshaping: The data set must be reshaped to list all spectra in a matrix. Create a two-dimensional matrix with all recorded spectra listed in the rows and columns. The data must be reshaped into a hypercube-formatted matrix to analyze the Raman image scans. This is a three-dimensional matrix with each x representing the x coordinate of the pixel, each y representing the y coordinate of the pixel, and the z value representing spectra.
5. Trimming: Spectral areas that contain no information (before 400 cm^{-1}, between 1800 and 2800 cm^{-1}, and after 3140 cm^{-1}) should be eliminated by cropping the data set to reduce the total data size and enable faster processing.
6. Normalization: Normalization is performed by z-scoring, setting the mean of the data to 0 and the standard deviation to 1 to compare the different data sets (*see* **Note 8**).
7. Smoothing: The data can be smoothed using the Savitzky–Golay filter by adjusting the polynomial filter and frame length depending on the data.

Spectral comparison and statistical analysis

1. Compare endosymbiont spectra with reference spectra (e.g., standard biomolecules).
2. Identify the biomarkers unique to symbionts.
3. Apply principal component analysis (PCA) to identify variations in spectral data.
4. Use clustering methods (e.g., hierarchical clustering or k-means) to distinguish between endosymbiont populations.

4 Notes

1. Avoid over-fixation; otherwise, it interferes with Raman signals.
2. It is highly recommended to observe the cell suspension that should be in the active phase.
3. Attention should be paid to achieve spatial separation of cells and to limit the aggregation of cells.
4. The choice depends on the sample and expected Raman bands to avoid fluorescence interference. The 532 nm lasers offer the highest signal intensity but will have a higher fluorescent background compared to 785 nm.
5. For most biological samples, the parameters that produce good results are as follows: laser: 532 nm; laser power: 20–40 mW; accumulation time: 0.2–0.5 s.
6. It is advisable to observe the prepared slides using a white-light stereomicroscope and the collection of images from single spots on the plate revealing the spatial distribution of the cells.
7. Eliminate peak with tenfold higher intensities as the mean at a given wavelength.
8. Depending upon the type of data, other normalization types might be better suitable, for example, Pareto scaling.

References

1. Chandarana KA, Gohil K, Dwivedi MK, Amaresan N (2023) Culture-independent and culture-dependent approaches in symbiont analysis. In: Microbial symbionts. Academic Press, Amsterdam, pp 723–742
2. Dodo K, Fujita K, Sodeoka M (2022) Raman spectroscopy for chemical biology research. J Am Chem Soc 43:19651–19667
3. Yang Q, Cahn JK, Piel J, Song YF, Zhang W, Lin HW (2022) Marine sponge endosymbionts: structural and functional specificity of the microbiome within Euryspongia arenaria cells. Microbiol Spectr 3:e02296-21
4. Rebrosova K, Samek O, Kizovsky M, Bernatova S, Hola V, Ruzicka F (2022) Raman spectroscopy—a novel method for identification and characterization of microbes on a single-cell level in clinical settings. Front Cell Infect Microbiol 12:866463
5. Haessler A, Jung N, Windbergs M (2023) Unraveling molecular composition in biological samples–systematic evaluation of statistical methods for the analysis of hyperspectral Raman data. Anal Chem 95:17646–17653

Chapter 16

Next-Generation Sequencing Approaches to Study Endosymbionts

Jinal Naik and Natarajan Amaresan

Abstract

Endosymbionts are types of microorganisms that reside within the cells of a host and play a vital role in host biology, including nutrition, defense, and development. Some endosymbionts can be found either inside cells or attached to the surface of cells. Metagenomics is a recently used technique for the analysis of environmental microorganisms. Metagenomics analysis and next-generation sequencing played important roles in the identification and characterization of endosymbionts. This chapter elaborates on endosymbiont isolation, purification, and amplification of genome sequencing. Metagenomics sequencing analysis was performed for the identification of endosymbiont.

Key words Endosymbiont, Metagenomics analysis, Next-generation sequencing

1 Introduction

An organism that lives within the cells or body of another organism and develops a mutualistic relationship is known as an endosymbiont. The relationship between them is beneficial for both. Such organisms can live either intracellularly or extracellularly. Endosymbionts are classified as primary and secondary endosymbionts. The endosymbiont associated with host by a long obligate relationships are known as primary endosymbiont. In contrast, secondary endosymbionts are more recent associates. They are not obligate, and sometimes they may be horizontally transferred between hosts and may even live outside the insect host [1]. Earlier traditional endosymbiont methods include fluorescent in situ hybridization (FISH), transmission electron microscopy (TEM), and 16 s rRNA sequencing [2]. In recent eras, the next-generation sequencing method has overcome the use of such traditional methods due to its advanced and faster mechanisms. Genome research is strongly

N. Amaresan and Sapna Chandwani (eds.), *Endosymbionts*, Springer Protocols Handbooks,
https://doi.org/10.1007/978-1-0716-4771-4_16,

supported by the NGS approach which provides higher-throughput data at lower cost. Such relationships contribute to the dominance of insects [3].

1.1 Metagenomics Approach to Study Endosymbionts

The metagenomics research approach is initiated with the biological environment which is associated with biological important varieties and microbial communities and plays a gentle role in medical, agriculture, and environmental fields. It is the culture-independent analysis of the total genomic content of microorganisms associated with any environment. The application of metagenomics has played a promising role in the case of medical and environmental microbiology. Metagenomics has special importance in soil microbiology [4].

1.2 Metaproteomics Approach to Study Endosymbionts

Genomic analysis expresses the importance of microbiome in the human health. In this research era, genomics gives limited information on the functional aspect of the microbiome However, the proteomic analysis provides brief functional information on host–microbe interaction [5].

Microbes possess variability based on their interacting individuals [6]. Such type of host–microbe interaction represents a symbiotic relationship.

1.3 Meta-transcriptomics Approach to Study Endosymbionts

Metatranscriptomics has emerged as a very useful technology for the study of microbiomes from RNA-seq reads. This method provides additional information compared to the sequencing of ribosomal genes because the gene expression can also be analyzed [7].

2 Materials

2.1 Metagenomics Approach to Study Endosymbionts

1. Endosymbiont sample.
2. Ethanol: 70% ethanol is used to purify the sample.
3. Quick-DNA MicroPrep kit.
4. *Nextera DNA Flex Library Prep kit*: It provides an automation-ready workflow process for Illumina sequencers.
5. Qubit dsDNA HT Assay kit.

2.1.1 DNA Isolation from Endosymbiont Using Lysis Buffer

1. Lysis buffer.
2. Specific primers.
3. Taq DNA polymerase.
4. DNTPS.
5. 10× buffer.

6. 100 μL of 1× PBS: For preparation of PBS, take 13.7 mM NaCl, 2.7 mM KCl, 10 mM Na_2HPO_4, and 1.8 mM KH_2PO_4.
7. DNA amplification kit.

2.2 Metaproteomics Approach to Study Endosymbionts

1. SDS lysis buffer: Add 10 mM Tris–HCl, followed by 2 mM EDTA, and at last add 1% SDS solution.
2. IAA solution: It is a synthetic plant growth regulator. Dissolved IAA powder in a solvent such as ethanol or water.
3. 50 mM of ammonium bicarbonate: Dissolve ammonium bicarbonate into Milli-Q water.

2.3 Meta-transcriptomics Approach to Study Endosymbionts

1. 770–1180-μm glassbeads.
2. miRNeasy MiniKit and purify RNA.

3 Methods

3.1 Metagenomics Approach to Study Endosymbionts

3.1.1 DNA Isolation from Endosymbiont

1. Collect the endosymbiont sample and wash it with ethanol to remove possible external contaminants (*see* **Note 1**).
2. For that, agitate the tube manually, and collect ethanol with a pipette using a thin pipette tip. Refill the tube with ethanol. Repeat the step three times.
3. Use Quick-DNA MicroPrep kit to isolate genomic DNA.
4. Quantify total DNA by using the Qubit High Sensitivity dsDNA assay.
5. Prepare the DNA library using the Nextera DNA Flex Library Prep kit.
6. For library preparation, cut DNA enzymatically and add adapters using a single step.
7. Quantify the prepared library using the Qubit dsDNA HT Assay kit.

3.1.2 DNA Isolation from Endosymbiont Using Lysis Buffer [8]

1. Collect the host sample species of endosymbiont with specific atmospheric conditions.
2. Homogenize endosymbiont carrying host sample using 30 μL of lysis buffer.
3. Incubate homogenate at 65 °C for 1 h and then 100 °C for 10 min.
4. Perform polymerase chain reaction (PCR) with isolated DNA using a specific primer.
5. Perform the PCR reactions in a final volume of 25 μL containing 1 U of Taq DNA polymerase, 2.5 μL 10× buffer, 0.2 mM

dNTP, 0.2 mM of each primer, and 2 μL of whitefly DNA (*see* **Note 2**).

6. Purify the PCR-amplified product using a DNA gel extraction kit with the recommended protocol.
7. Using a DNA sample sequencing system, perform the sequencing of the purified DNA.
8. Analyze the sequences using the Basic Local Alignment Search Tool (BLAST) to identify the cryptic species of whitefly and avoid contamination of other species.

3.1.3 Endosymbiont Identification and Localization

1. Perform PCR of the isolated DNA to amplify the specific genes of each endosymbiont within the host organisms.
2. Perform agarose gel electrophoresis (1% agarose gel, Tris–acetate–EDTA as running buffer, voltage 5 V/cm) using standard protocol to identify the specific gene.
3. Recover and sequence each band to determine different species within endosymbiont organisms.
4. Perform FISH (fluorescence in situ hybridization) analysis to identify the location of endosymbionts.

3.1.4 Endosymbiont Dissection and Purification

1. For dissection of endosymbiont, take 100 μL of 1× PBS solution onto a microscope slide. Immerse the host organism into PBS solution (*see* **Note 3**).
2. With the help of entomological needles, pull endosymbiont out from the host organism.
3. Make a hole on the side of the host organism and remove the endosymbiont by pressing the nymph on the other side.
4. Mount a 20-μL microloader on a 0.5–10-μL pipette. Extract the individual endosymbiont.
5. Remove the contamination by providing washing to endosymbiont. Repeat the step three times.
6. Pipette the washed endosymbiont immediately into a centrifuge tube containing 60 μL of 1× PBS solution.
7. For filtration, use a syringe filter of a specific filter membrane.

3.1.5 Amplification of Endosymbiont Genomes

1. Use a DNA amplification kit to amplify the filtrate by the recommended protocol with some modification.
2. For amplification, add 1.5 μL of filtrate to the centrifuge tube followed by 1.5 μL of the buffer D2 and mix well.
3. Incubate mixture on ice for 10 min and add 1.5 μL of stop solution.
4. Add 15 μL of the reaction buffer and 1 μL of the DNA polymerase and mix gently.

5. Incubate the mixture for 16 h, at 30 °C followed by 3 min at 65 °C.
6. Perform PCR directly on the amplified filtrate (dilute the amplified samples if necessary) by using specific primers designed for endosymbiont to confirm the bacterial species.
7. Conduct PCR to confirm whether there is contamination of the host genome by using the host gene with a specific primer.

3.1.6 Endosymbiont Metagenome Sequencing

Subject the amplified filtrate to a spectrophotometer and a fluorometer (according to the manufacturer's instructions) before sequencing to test the quality of samples and also to analyze whether the samples meet the criteria for sequencing or not.

3.2 Metaproteomics Approach to Study Endosymbionts

3.2.1 Preparation of Sample

1. Use SDT–lysis buffer sodium dodecyl sulfate (SDS) with 0.1 N dithiothreitol (DDT) to lysis the sample.
2. Add 1:10 (sample:buffer) to the sample pellets. Perform centrifugation at 21,000×*g* for 5 min in order to get the pellet of debris.
3. After pelleting, use the OMNI Bead Ruptor 24, then incubate at 95 °C for 10 min to ensure the cells are distributed within the lysing matrix tube.
4. For filter-aided sample preparation, mix 30 μL of the clear lysate with 200 μL of UA solution (8 M urea in 0.1 M Tris/HCl, pH 8.5) in a 10 kDa MWCO 500 μL centrifugal filter unit (VWR International) and centrifuge at 14,000×*g* for 40 min.
5. Add another 200 μL of UA solution to the filter unit and centrifuge at 14,000×*g* for 40 min.
6. Then, add 100 μL of IAA solution to the filter and incubate at 22 °C for 20 min (*see* **Note 4**).
7. Remove the IAA solution using centrifugation and wash the filter with 100 μL of UA three times followed by washing the filter using 50 mM ammonium bicarbonate.
8. Add 1 to 2 μg of MS grade trypsin in 40 μL of ABC-added filter and incubate it at 37 °C in a wet chamber overnight.
9. Elute out the peptide by centrifugation at 14,00×*g* for 20 min, followed by the addition of 50 μL of 0.5 M NaCl, and further centrifugation is followed.
10. Desalt the peptides using Sep-Pak C18 Plus Light Cartridges and C18 spin columns (Thermo Scientific Pierce) (*see* **Note 4**).

3.2.2 LC/MS Analysis

1D and 2D LC/MS analyses are performed using a prepared proteomic sample for the identification of peptides present in it and to identify appropriate protein values.

1. For one-dimensional (1D) LC/MS analysis, load 2 μg of peptides onto 5 mm, 300 μm ID C18. Acclaim pre-column using an UltiMate™ 3000 RSLCnano Liquid Chromatography (Thermo Fisher Scientific) with loading solvent A (2% acetonitrile, 0.05% trifluoroacetic acid (TFA)). Eluent A is 0.1% formic acid in water, and eluent B is 0.1% formic acid in acetonitrile.
2. The pre-column was switched in line with a 50-cm × 75-μm analytical EASY-Spray column packed with PepMap RSLC C18, 2 μm material (Thermo Fisher Scientific), which is heated to 45 °C.
3. Separate peptides using an analytical column using 260- or 460-min gradients, and mass spectra are acquired [9].
4. Use a 260-min gradient (from 2% B to 31% B in 200 min, increased to 50% B in the next 40 min, followed by a 20-min wash at 99% B) and a 460-min gradient (from 2% B to 31% B in 363 min, increased to 50% B in 70 min, followed by a 27 min wash at 99% B) for 1D LC.
5. For 2D LC/MS analysis, load 11 μg of peptide onto a 10-cm, 300-μm ID Poros 10 S SCX column (Thermo Fisher Scientific) using the UltiMate™ 3000 RSLCnano LC with loading solvent B (2% acetonitrile, 0.5% formic acid).
6. Elute peptide using SCX column onto the C18 pre-column using 20 μL injection of salt plugs from the autosampler with increasing concentrations (12 salt plugs, 0–2000 mM NaCl).
7. After each salt plug injection, switch the pre-column in line with the 50-cm × 75-μm analytical EASY-Spray column and separate the peptides using a 120-min gradient (from 2% B to 31% B in 82 min, increased to 50% B in 10 min, held at 99% B for 9 min, and re-equilibrated at 2% B for 19 min).

3.2.3 Data Availability

Deposit the mass spectrometry metaproteomics data and protein sequence databases to the ProteomeXchange Consortium via the PRIDE. Mass spectrometry and metaproteomics data is verified with a protein sequence database using Pride software (*see* **Note 5**).

3.3 Meta-transcriptomics Approach to Study Endosymbionts

3.3.1 Sample Collection

Species having endosymbiont relation is collected and identified using basic characteristics. Store the endosymbiont or their tissue for later extraction of RNA.

3.3.2 Symbiont RNA Extraction and Enrichment

1. Homogenize gill tissue of symbiont using FastPrep FP120 tissue homogenizer with 770–1180-μm glassbeads.
2. Extract total gill RNA with chloroform–ethanol method by the use of miRNeasy MiniKit and purify RNA.

3. MICROENRICH kit is used for the enrichment of symbiont RNA extract.
4. The kit can remove Eukaryotic 18S and 28S rRNA.

3.3.3 RNA Amplification and cDNA Synthesis

1. Extract the linearly amplified RNA and convert it into cDNA [10].
2. RNA is poly-adenylated using the MessageAmp™ II-Bacteria kit and converted to double-stranded cDNA via reverse transcription.
3. Transcribe antisense RNA to produce large quantities of single-stranded RNA.
4. Convert amplified RNA into double-stranded cDNA using the SuperScript® III.
5. Purify CDNA with the QIA quick PCR purification kit and remove the poly-A tail from the sequence.
6. Purified cDNA is used for identification using sequencing analysis.

3.3.4 Sequencing

1. Generate a single-stranded DNA library with purified CDNA and emulsion PCR (using standard PCR protocol).
2. Sequence the amplified library fragment using a full plate run on genome sequencer.

3.3.5 Data Analysis

1. Compress sequence data by removing duplicate sequencing using the program CD-HIT [11].
2. Identify non-duplicate reads matching by BLAST searches against a custom database of prokaryotic and eukaryotic small- and large-subunit rRNA nucleotide sequences.
3. Identify protein-coding gene sequences by BLAST X analysis.
4. Identify the taxonomy of protein-coding genes using MEGAN [12].

4 Notes

1. Ensure ethanol is completely removed after washing steps to prevent interference with downstream reactions.
2. Prepare PCR mixes on ice to reduce non-specific amplification.
3. Perform dissections under sterile PBS to avoid drying or contamination.
4. Avoid introducing air bubbles during filtration and peptide desalting.
5. Use proper software versions and databases (BLAST, MEGAN, PRIDE) to avoid misannotation.

References

1. Harish ER, Chellappan M, MakeshKumar T, Mathew D, Ranjith MT, Girija D (2019) Next-generation sequencing reveals endosymbiont variability in cassava whitefly, Bemisia tabaci, across the agro-ecological zones of Kerala, India. Genome 62:571–584
2. Iovieno A, Ledee DR, Miller D, Alfonso EC (2010) Detection of bacterial endosymbionts in clinical Acanthamoeba isolates. Ophthalmology 117:445–452
3. Su Q, Oliver KM, Pan H, Jiao X, Liu B, Xie W et al (2013) Facultative symbiont Hamiltonella confers benefits to Bemisia tabaci (Hemiptera: Aleyrodidae), an invasive agricultural pest worldwide. Environ Entomol 42:1265–1271
4. Hozzein WN (ed) (2020) Metagenomics: basics, methods and applications. BoD–Books on Demand, Norderstedt
5. Starr AE, Deeke SA, Li L, Zhang X, Daoud R, Ryan J, Ning Z, Cheng K, Nguyen LV, Abou-Samra E, Lavallée-Adam M (2018) Proteomic and metaproteomic approaches to understand host–microbe interactions. Anal Chem 90:86–109
6. Costello EK, Lauber CL, Hamady M, Fierer N, Gordon JI, Knight R (2009) Bacterial community variation in human body habitats across space and time. Science 326:1694–1697
7. Stewart FJ, Dmytrenko O, DeLong EF, Cavanaugh CM (2011) Metatranscriptomic analysis of sulfur oxidation genes in the endosymbiont of Solemya velum. Front Microbiol 2:134
8. Kleiner M, Thorson E, Sharp CE, Dong X, Liu D, Li C, Strous M (2017) Assessing species biomass contributions in microbial communities via metaproteomics. Nat Commun 8:1558
9. Petersen JM et al (2016) Chemosynthetic symbionts of marine invertebrate animals are capable of nitrogen fixation. Nat Microbiol 2:16195
10. Frias-Lopez J, Shi Y, Tyson GW, Coleman ML, Schuster SC, Chisholm SW, DeLong EF (2008) Microbial community gene expression in ocean surface waters. Proc Natl Acad Sci 105:3805–3810
11. Li W, Godzik A (2006) Cd-hit: a fast program for clustering and comparing large sets of protein or nucleotide sequences. Bioinformatics 22:1658–1659
12. Huson DH, Auch AF, Qi J, Schuster SC (2007) MEGAN analysis of metagenomic data. Genome Res 17:377–386

Part III

Endosymbionts Having Potential for Biologically Active Metabolites

Chapter 17

Screening of Endosymbionts for Plant Growth–Promoting Metabolites

Pfariso Maumela and Mahloro H. Serepa-Dlamini

Abstract

This chapter outlines the biochemical and metabolomic methods used to screen endosymbionts for plant growth–promoting metabolites. It will cover biochemical techniques for detecting the synthesis of siderophores, indole acetic acid, and 1-aminocyclopropane-1-carboxylate deaminase. Additionally, the chapter will explore the use of non-targeted metabolomics and liquid chromatography–mass spectrometry to identify metabolites that promote plant growth.

Key words Siderophores, Metabolomics, Spectrometry, Phytohormones, Endosymbionts

1 Introduction

Endosymbiotic microbes colonize the extracellular and intracellular spaces of their plant hosts to establish mutualistic relationships, thereby promoting plant growth and development [1]. Endosymbionts promote plant growth and development through mechanisms including enhancement of nutrient availability, regulation of phytohormones, and improvement of stress tolerance [2]. Plant growth–promoting (PGP) endosymbionts enhance nutrient availability through nitrogen fixation and the solubilization of inorganic phosphates [3].

Phosphorus is an essential plant macronutrient required for the synthesis of phospholipids and adenosine triphosphate [4]. Phosphorus is thus involved in key plant functions including energy transfer, photosynthesis, nutrient metabolism, and transport [5]. However, a significant amount of phosphorus, from the soil, is available as insoluble inorganic salts [6]. Endosymbionts are therefore, capable of synthesizing and secreting phosphate solubilizing enzymes which transform phosphate salts into monobasic or dibasic phosphates which are easily absorbed by the plants [3].

N. Amaresan and Sapna Chandwani (eds.), *Endosymbionts*, Springer Protocols Handbooks,
https://doi.org/10.1007/978-1-0716-4771-4_17,

Plant growth–promoting endosymbionts synthesize siderophores which are chelating agents with a high specificity for binding iron [1]. Siderophores enhance the host plant's ability to overcome metal stress and iron deficiency [7]. Petrobactin is a siderophore important for iron acquisition through membrane-associated substrate-binding proteins and ABC transporters [8]. Siderophores can also form stable complexes with other heavy metals such as Cd, Pb, Zn, and Cu [9], thus protecting the host plant against heavy metal stress.

Endosymbionts are involved in phytohormone regulation directly or indirectly [10]. Indoleacetic acid (IAA), an auxin, is a plant phytohormone that is also synthesized by endosymbionts [11]. IAA is a PGP factor, which affects cell division and stimulates plant cell elongation even under stress [12]. Endosymbionts are indirectly involved in phytohormone regulation and plant growth through the synthesis of 1-aminocyclopropane-1-carboxylate (ACC). ACC is a PGP factor important in the reduction of ethylene, a stress-generating hydrocarbon, which subsequently promotes plant growth [13].

Biochemical tests provide a convenient and rapid approach for the identification and screening of microbes [14]. Biochemical tests are based on the metabolic activities of microbes, such as nutrient utilization, secondary metabolite synthesis, and enzyme production or utilization [15]. Mass spectrometry can be used to augment biochemical tests, due to its ability to analyze and identify a wide range of compounds, even in small concentrations [16]. This chapter, thus, discusses the basic techniques and corresponding instrumentations of the biochemical tests and liquid chromatography–mass spectrometry (LC–MS) non-targeted metabolomics approach for the screening of endosymbionts for plant growth–promoting metabolites. The identification of metabolites is essential and a preliminary step in elucidating the structure and chemical properties of the metabolites and consequently their mechanisms of action.

2 Materials

Clean all glassware with 6 M HCl. Prepare all solutions with double-distilled water.

2.1 CAS Agar

1. CAS solution: Dissolve 60.5 mg CAS in 50 mL water.
2. Iron III solution: 1 mM $FeCl_3{\cdot}6H_20$, 10 mM HCl. Dissolve 2.7 mg $FeCl_3{\cdot}6H_20$ and 3.65 mg HCl in 10 mL of water.
3. HDTMA solution: Dissolve 72.9 mg in 40 mL water.
4. 10× MM9 salts: Dissolve 6 g Na_2HPO_4, 3 g KH_2PO_4, 1 g NH_4Cl, and 0.5 g NaCl in 100 mL water.

5. Glucose stock solution: 20% w/v. Dissolve 20 g in 100 mL of water.
6. NaOH stock solution: Dissolve 25 g of NaOH in 25 mL water.
7. Casamino acid solution: 10% w/v. Dissolve 3 g of casamino acid in 30 mL of water.

2.2 DF Plates

1. Dworkin and Foster [17] minimal salts medium. Salts per L: 4.0 g KH_2PO_4, 6.0 g Na_2HPO_4, 0.2 g $MgSO_4{\cdot}7H_2O$, 2.0 g glucose, 2.0 g gluconic acid, and 2.0 g citric acid with trace elements: 1 mg $FeSO_4{\cdot}7H_2O$, 10 mg H_3BO_3, 11.2 mg $MnSO_4{\cdot}H_2O$, 124.6 mg $ZnSO_4{\cdot}7H_2O$, 78.22 mg $CuSO_4{\cdot}5H_2O$, 10 mg MoO_3, pH 7.2.

2.3 Salwoski Reagent

1. Salkowski reagent: 50 mL of 35% $HClO_4$ and 1 mL of 0.5 M $FeCl_3$.

2.4 Pikovskaya's Agar

1. Pikovskaya's medium agar [18]: nutrients per L: 0.5 g yeast extract, 10 g dextrose, 5 g $Ca_3(PO_4)_2$, 0.5 g $(NH_4)_2SO_4$, 0.2 KCl, 0.1 g $MgSO_4$, 0.0001 g $MnSO_4$, 0.0001 $FeSO_4$, and 15 g agar (*see* **Note 1**).

2.5 LC–MS

1. Liquid chromatography–quadrupole time-of-flight tandem mass spectrometer (LC–MS-9030 q-TOF, Shimadzu Corporation, Kyoto, Japan).
2. Shim-pack Velox C18 column (100 mm × 2.1 mm with particle size of 2.7 m).
3. Eluent A: 0.1% v/v formic acid in water.
4. Eluent B: 0.1% v/v formic acid in acetonitrile.

3 Methods

3.1 Siderophore Production

1. Mix 50 mL of the CAS solution with 10 mL of the iron III solution.
2. Add the mixture to the HDTMA solution while stirring.
3. Autoclave the resulting blue dye solution.
4. Mix and autoclave 750 mL of water, 100 mL of 10× MM9 salts, 15 g agar, 30.24 g PIPES (*see* **Note 2**), and 12 g of 50% (w/w) NaOH (*see* **Note 3**).
5. Cool the MM9/PIPES mixture to 50 °C. Add 30 mL of sterile casamino acid solution and 10 mL of sterile 20% glucose solution to the mixture (*see* **Note 4**).
6. Slowly add 100 mL of the blue dye solution along the glass wall with enough agitation to mix thoroughly and avoid foaming.

7. Aseptically pour plates.
8. The microbial strain is spot inoculated in the center of a CAS agar plate and incubated at 28 °C for 3 days [19]. An orange halo zone [20] around the colony indicates evidence of siderophore production.

3.2 ACC Deaminase

1. Prepare 1 L of DF minimal salts medium. Add 2 g of agar and 303 mg ACC.
2. Autoclave the media. Cool the prepared agar media and aseptically pour plates.
3. Spot inoculate the microbial strain and incubate at 28 °C for 3 days. Colony growth on ACC plates represents a positive outcome for ACC deaminase production (*see* **Note 5**).

3.3 IAA Production

1. Culture strain in broth (*see* **Note 6**) supplemented with 0.5 mg/mL of **L**-tryptophan and incubate the mixture for 4 days at 30 °C.
2. Centrifuge the culture broth at 10,000 rpm for 10 min. Mix 2 mL of the supernatant with 4 mL of Salkowski reagent [21]. Allow mixture to stand in the dark for 30 min at 25 °C. The production of a pink color from the reaction is a positive indicator of IAA production [22].

3.4 Phosphate Solubilization

1. Suspend the Pikovskaya's agar components in 1 L of water and boil to dissolve. Autoclave to sterilize the media.
2. Aseptically pour the plates.
3. Spot inoculate the strain in the middle of Pikovskaya's agar and incubate at 35 °C for 2 days (*see* **Note 7**). A halo or yellow zone around the colony is indicative of phosphate solubilization activity.

3.5 LC–MS

1. Filter sterilize the culture from the microbial strain with a 0.45-μm syringe filter and transfer it into a glass vial for injection into the LC–MS equipment.
2. Maintain the column oven temperature at 50 °C.
3. Use an injection volume of 5 μL and analytically separate the samples over a 30-min binary gradient.
4. Apply a constant flow rate of 0.04 mL/min using a binary solvent mixture of water with 0.1% formic acid and 0.1% formic acid in acetonitrile (*see* **Note 8**).
5. Keep eluent B at 5% from 0 to 3 min, gradually increase from 5% to 40% between 3 and 5 min, and finally increase to 40–95% between 5 and 23 min.
6. Keep eluent B isocratic at 95% between 23 and 25 min.

7. Return the gradient to the original conditions of 5% at 25–27 min, and re-equilibrate at 5% at 27–30 min.
8. Subject the liquid chromatographic eluents to a quadruple time-of-flight high-definition mass spectrometer for analysis in positive electrospray ionization (ESI) mode with the following conditions: 400 °C heat block temperature, 250 °C desolvation line (DL) temperature, 42 °C flight tube temperature, and 3 L/min nebulization and dry gas flow.
9. The data is acquired using the data-dependent acquisition (DDA) mode, which simultaneously generates MS1 and MS2 data for all ions within a mass-to-charge ratio (m/z) range of 100–1500 (precursor m/z isolation window) and an intensity threshold above 5000 (*see* **Note 9**).
10. Annotate metabolites (*see* **Notes 10, 11** and **12**).

4 Notes

1. Pikovskaya's agar is also commercially available. The prepared media should be stored below 8 °C and away from direct light.
2. PIPES will not dissolve below a pH of 5. Bring pH up to 6 and slowly add PIPES while stirring. The pH will drop as PIPES dissolves. While stirring, slowly bring the pH up to 6.8. Do not exceed 6.8 as this will turn the solution green.
3. NaOH is important to raise the pH to the pK_a of Pipes (6.8).
4. *E. coli*–specific ingredients: 10 mL glucose (20%) as a carbon source, 1 mL thiamine·HCl (0.2%), and 3 mL L-tryptophan (1%). *R. meliloti*–specific ingredients: 10 mL glucose or sucrose (20%) as a carbon source, 5 mL L-glutamic acid (10%, neutralized), and 2.5 mL (+)-biotin (0.02%).
5. Plates containing only DF minimal salts medium without ACC are used as a negative control, and those with $(NH_4)_2SO_4$ (0.2% w/v) are used as the positive control [23].
6. Use Luria broth for bacteria, potato dextrose broth for fungi, and yeast extract peptone dextrose broth for yeast [24].
7. Alternatively, incubate at 28 °C for 7 days.
8. The gradient technique is gradually increased from 3 to 30 min to facilitate the separation of the compounds within the samples.
9. The MS2 experiments are conducted utilizing argon gas as the collision gas and a collision energy of 35 eV with a spread of 5 and sodium iodide as a calibration solution to monitor high mass precision.

10. Metabolite annotation is completed at Metabolomics Standards Initiative (MSI) levels 2 and 3. The former is based on the retention time, mass-to-charge ratio (m/z), and fragmentation patterns matching data available from the databases in Sirius [25–27]. The fragments with no matches to anything on the databases are classified according to their compound class according to the molecular networking from Canopus on Sirius [26, 27].
11. Databases such as Metlin and PubChem can be subsequently used to identify the biological activities of the annotated metabolites [28].
12. Gas chromatography–mass spectrometry (GC–MS) can also be used for terpene-based metabolites [29].

Acknowledgement

This work was supported by the ZAR-NRF Thuthuka grant no. TTK210216586709 and Postdoctoral fellowship grant no. PSTD2204082662.

References

1. Kumar M, Ahmad S, Singh RP (2022) Plant growth promoting microbes: diverse roles for sustainable and ecofriendly agriculture. Energy Nexus 7:100133
2. Lyu D, Backer R, Berrué F, Martinez-Farina C, Hui JP, Smith DL (2023) Plant growth-promoting rhizobacteria (PGPR) with microbial growth broth improve biomass and secondary metabolite accumulation of Cannabis sativa L. J Agric Food Chem 71(19): 7268–7277
3. Sati D, Pandey SC, Pande V, Upreti S, Gouri V, Joshi T, Gangola S, Debbarma P, Pandey A, Samant M (2020) Toward an enhanced understanding of plant growth promoting microbes for sustainable agriculture. In: Recent advancements in microbial diversity. Academic Press, Oxford, pp 87–112
4. Tariq A, Zeng F, Graciano C, Ullah A, Sadia S, Ahmed Z, Murtaza G, Ismoilov K, Zhang Z (2023) Regulation of metabolites by nutrients in plants. In: Plant ionomics: sensing, signaling, and regulation, vol 3. Wiley, Hoboken, pp 1–8
5. Wang Y, Chen YF, Wu WH (2021) Potassium and phosphorus transport and signaling in plants. J Integr Plant Biol 63(1):34–52
6. Goswami SP, Maurya BR, Dubey AN, Singh NK (2019) Role of phosphorus solubilizing microorganisms and dissolution of insoluble phosphorus in soil. Int J Chem Stud 7(3): 3905–3913
7. Gupta R, Khan F, Alqahtani FM, Hashem M, Ahmad F (2024) Plant growth–promoting Rhizobacteria (PGPR) assisted bioremediation of heavy metal toxicity. Appl Biochem Biotechnol 196(5):2928–2956
8. Zawadzka AM, Abergel RJ, Nichiporuk R, Andersen UN, Raymond KN (2009) Siderophore-mediated iron acquisition systems in Bacillus cereus: identification of receptors for anthrax virulence-associated petrobactin. Biochemist 48(16):3645–3657
9. Dimkpa CO, Svatoš A, Dabrowska P, Schmidt A, Boland W, Kothe E (2008) Involvement of siderophores in the reduction of metal-induced inhibition of auxin synthesis in Streptomyces spp. Chemosphere 74(1): 19–25
10. Sarkar S, Dey A, Kumar V, Batiha GE, El-Esawi MA, Tomczyk M, Ray P (2021) Fungal endophyte: an interactive endosymbiont with the capability of modulating host physiology in myriad ways. Front Plant Sci 12:701800
11. Soundar Raju C, Aslam A, Thangadurai D, Sangeetha J, Kathiravan K, Shajahan A (2020) Indole acetic acid (IAA) producing endophytic bacteria on direct somatic embryogenesis and

plant regeneration of Exacum travancoricum Bedd. Vegetos 33:690–702

12. Waqas M, Khan AL, Kamran M, Hamayun M, Kang SM, Kim YH, Lee IJ (2012) Endophytic fungi produce gibberellins and indoleacetic acid and promotes host-plant growth during stress. Molecules 17(9):10754–10773
13. del Carmen O-MM, Santoyo G (2021) Plant-microbial endophytes interactions: scrutinizing their beneficial mechanisms from genomic explorations. Curr Plant Biol 25:100189
14. Peng X, Ren B, Li Y, Zhou X, Xie J, Zhou C, Zhang D, Zheng X, Zhou X (2020) Techniques for oral microbiology. In: Zhou X, Li Y (eds) Atlas of Oral microbiology: from healthy microflora to disease. Springer, Singapore. https://doi.org/10.1007/978-981-15-7899-1_2
15. Tamang JP (2014) Biochemical and modern identification techniques: microfloras of fermented foods. Academic Press, Oxford
16. Thomas SN, French D, Jannetto PJ, Rappold BA, Clarke WA (2022) Liquid chromatography–tandem mass spectrometry for clinical diagnostics. Nat Rev Methods Primers 2(1):96
17. Dworkin M, Foster J (1958) Experiments with some microorganisms which utilize ethane and hydrogen. J Bacteriol 75(5):592–603
18. Pikovskaya R (1948) Mobilization of phosphorus in soil in connection with vital activity of some microbial species Mikrobiologiya 17: 362–370.
19. Schwyn B, Neilands J (1987) Universal chemical assay for the detection and determination of siderophores. Anal Biochem 160(1):47–56
20. Ames-Gottfred NP, Christie BR, Jordan DC (1989) Use of the chrome azurol S agar plate technique to differentiate strains and field isolates of rhizobium leguminosarum biovar trifolii. Appl Environ Microbiol 55(3):707–710
21. Gordon SA, Weber RP (1951) Colorimetric estimation of indoleacetic acid. Plant Physiol 26(1):192
22. Zhang BX, Li PS, Wang YY, Wang JJ, Liu XL, Wang XY, Hu XM (2021) Characterization and synthesis of indole-3-acetic acid in plant growth promoting Enterobacter sp. RSC Adv 11(50):31601–31607
23. Ali SZ, Sandhya V, Venkateswar Rao L (2014) Isolation and characterization of drought-tolerant ACC deaminase and exopolysaccharide-producing fluorescent Pseudomonas sp. Ann Microbiol 64:493–502
24. Carvajal M, Godoy L, Gebauer M, Catrileo D, Albornoz F (2024) Screening for indole-3-acetic acid synthesis and 1-aminocyclopropane-carboxylate deaminase activity in soil yeasts from Chile uncovers Solicoccozyma aeria as an effective plant growth promoter. Plant Soil 496(1):83–93
25. Djoumbou Feunang Y, Eisner R, Knox C, Chepelev L, Hastings J, Owen G, Fahy E, Steinbeck C, Subramanian S, Bolton E, Greiner R (2016) ClassyFire: automated chemical classification with a comprehensive, compul0e taxonomy. J Cheminf 8:1–20
26. Kim HW, Wang M, Leber CA, Nothias LF, Reher R, Kang KB, Van Der Hooft JJ, Dorrestein PC, Gerwick WH, Cottrell GW (2021) NPClassifier: a deep neural network-based structural classification tool for natural products. J Nat Prod 84(11):2795–2807
27. Sumner LW, Amberg A, Barrett D, Beale MH, Beger R, Daykin CA, Fan TW, Fiehn O, Goodacre R, Griffin JL, Hankemeier T (2007) Proposed minimum reporting standards for chemical analysis: chemical analysis working group (CAWG) metabolomics standards initiative (MSI). Metabolomics 3:211–221
28. Serralheiro ML, Guedes R, Fadel SR, Bendif H (2020) Data on identification of primary and secondary metabolites in aqueous extract of Verbascum betonicifolium. Data Brief 32: 106146–106157
29. Xavier V, Spréa R, Finimundy TC, Heleno SA, Amaral JS, Barros L, Ferreira IC (2023) Terpenes. In Natural secondary metabolites: from nature, through science, to industry. Cham: Springer International Publishing.

Chapter 18

Screening of Endosymbionts for EPS Production for Enhancement of Symbiotic Proficiency with the Host

Bhavana V. Mohite, Sunil H. Koli, Pradnya B. Nikam, Narendra S. Salunkhe, Jitendra D. Salunkhe, and Satish V. Patil

Abstract

The microbial loop is a central system in the rhizosphere, where most microbes compete, and their various activities establish a distinct soil microflora. These microorganisms include bacteria, fungi, protozoa, viruses, and mold which may produce antibiotics, exopolysaccharide (EPS), and organic acid or act as bacterivorous organisms. Endosymbionts are organisms that live inside a host in a symbiotic relationship without causing harm. The biologically active compounds or metabolites produced by endosymbionts benefit the host. EPSs are high molecular weight polymeric molecules composed of sugar residues. The characteristic properties of EPS such as high water holding capacity, protection against biotic and abiotic stress, and nutrient acquisition benefit both the producing endosymbionts and their host. The ability of endosymbionts to produce EPS enhances their symbiotic association with the host. EPS producers thus play an important role in soil fertility and crop sustainability. The present protocol briefly describes methods for screening EPS-producing endosymbionts.

Key words Exopolysaccharides, Symbiosis, Negative staining, Methylene blue, Biofilm

1 Introduction

Endosymbiosis is the ability of an organism to live inside a host in a symbiotic association without any harm and often benefits the host. Endophytes are a significant source of unique metabolites and potential bioactive compounds [1]. The horizontal gene transfer among the host and its endosymbionts can enable the synthesis of metabolites similar or identical to those produced by the host itself [2].

Exopolysaccharides (EPSs) are polymeric metabolic products synthesized and secreted surrounding the bacterial cell surface [3]. EPSs consist of various organic and inorganic substances, among which carbohydrates are the prime constituent. Their composition and structure are widely variable; EPS may be

N. Amaresan and Sapna Chandwani (eds.), *Endosymbionts*, Springer Protocols Handbooks,
https://doi.org/10.1007/978-1-0716-4771-4_18,

homopolysaccharides, such as dextran, mutans, and levans, or heteropolysaccharides. Higher concentrations of EPS are typically obtained when the carbon to nitrogen (C:N) ratio in the medium is elevated [4]. Several studies have shown that EPS production increases under stress conditions such as salinity, low temperature, heavy metal exposure, and radiation [5, 6]. The biosynthesis of diverse EPS types involves multiple genes, and their expression can modulated in response to varying environmental conditions [7].

The role of EPS in host–symbiont interactions includes (a) protection of bacteria against environmental stresses by biofilm formation, (b) functioning as signaling molecules involved in host specificity, (c) facilitating attachment and successful interaction during nodulation, and (d) evasion of plant defense responses, including protection against plant-derived antimicrobial compounds [8].

(a) Under stressful or harsh environmental conditions, aggregation and biofilm formation enhance bacterial survival. EPS produced during biofilm formation protects free-living symbionts by shielding them from the surrounding environment [9]. The production of EPS has long been suggested to be involved in the tolerance of various stresses encountered by bacteria. Generally, rhizobial strains that produce higher amounts of EPS exhibit greater tolerance to adverse conditions compared to those producing lower amounts. Additionally, oxidative stress is known to induce EPS production, which plays a vital role in the successful establishment of symbiosis [10].

(b) The composition of EPS produced by symbiotic bacteria also acts as a signaling molecule, contributing to host plant specificity. For instance, *Rhizobial lipochitin* oligosaccharides (Nod factors), bacterial cell-surface components, and low molecular weight metabolites function in host recognition and signaling pathways [11].

(c) Attachment and successful interaction in nodulation: EPS plays a key role in the elongation of the infection thread by masking cellulose fibrils, preventing bacterial autoagglutination, and stimulating colonization of the infection thread [12].

(d) Avoidance of plant defense responses and protection against plant antimicrobial compounds: EPS produced during the early stages of infection acts as a diffusion barrier, protecting bacteria against toxic H_2O_2 generated by the host [13]. Structural modifications in specific surface polysaccharides can lead to increased sensitivity to host antimicrobial compounds. EPS may also contribute by masking key bacterial antigens during infection of nodule cells [14].

Considering the vital roles of EPS in enhancing symbiotic proficiency, this protocol outlines methods for screening EPS-producing endosymbionts. The screening of various types of EPS from different endosymbiotic bacteria is emphasized for their specific functional advantages in promoting effective symbiosis with their respective host plants.

2 Materials

2.1 Collection of Sample and Preparation of Extract

1. Disease-free wound-free plant material (leaves, roots, nodules).
2. 70% ethanol and 2.5% sodium hypochlorite.
3. Sterile saline, mortar pestle.
4. Sterile tubes with 2–5 mL saline solution.

2.2 Screening Assay

1. Exopolysaccharide screening agar medium 1 (*see* **Note 1**).
2. EPS screening medium 2. Low-strength nutrient agar (*see* **Note 2**).
3. YEMA agar with Congo red dye (*see* **Note 3**).
4. Nitrogen-free Burk's medium (*see* **Note 4**).
5. HS medium (*see* **Note 5**).
6. Nigrosine or Indian ink.
7. Glass slides.
8. Propanol.
9. Separating funnel.
10. PDA agar plates (*see* **Note 6**).

3 Method

3.1 Preparations of Sample

1. Collect the plant material (roots, shoots, leaves, nodules).
2. Disinfect the surface with 70% ethanol for 60 s, then rinse with sterile water.
3. Disinfect again with 2.5% sodium hypochlorite for 5 min, followed by a rinse with sterile water.
4. Repeat disinfection with 70% alcohol for 30 s.
5. Aseptically cut the material into small pieces and transfer them into sterile 0.8% saline.

6. Crush the sample in 0.8% sterile saline using a sterile glass rod. Incubate on a shaker at 120 rpm for 2 h and use the resulting suspension for isolation of EPS-producing endosymbionts.

3.2 Isolation of EPS-Producing Bacterial Endosymbionts

1. Streak the sample onto EPS screening medium 1 for the isolation of endosymbiotic bacteria from leaves, stems, and roots (*see* **Note 1**).
2. For the isolation of EPS-producing *Rhizobium* endosymbionts, crush the surface-sterilized nodule in saline and streak the suspension onto modified YEMA agar (*see* **Note 3**).
3. Similarly, streak the root nodule suspension onto EPS screening medium 2 (*see* **Note 2**).
4. Incubate the plates at 30 °C for 48 h, and observe the colony characteristics.

After incubation, select sticky, mucoid colonies and those showing zones of dye accumulation. Subculture these bacteria onto the same medium for purification. Dye accumulation is observed surrounding EPS-producing bacterial cells on glutamate-containing media with dyes such as neutral red (Fig. 1).

3.3 Screening of EPS-Producing Endosymbionts

1. Select a sticky, mucoid colony and prepare a suspension in 0.85% physiological saline.
2. Place a drop of 10% aqueous solution of nigrosine or India ink on one end of the glass slide, and mix it with a drop of the bacterial suspension using a sterile nichrome wire loop.

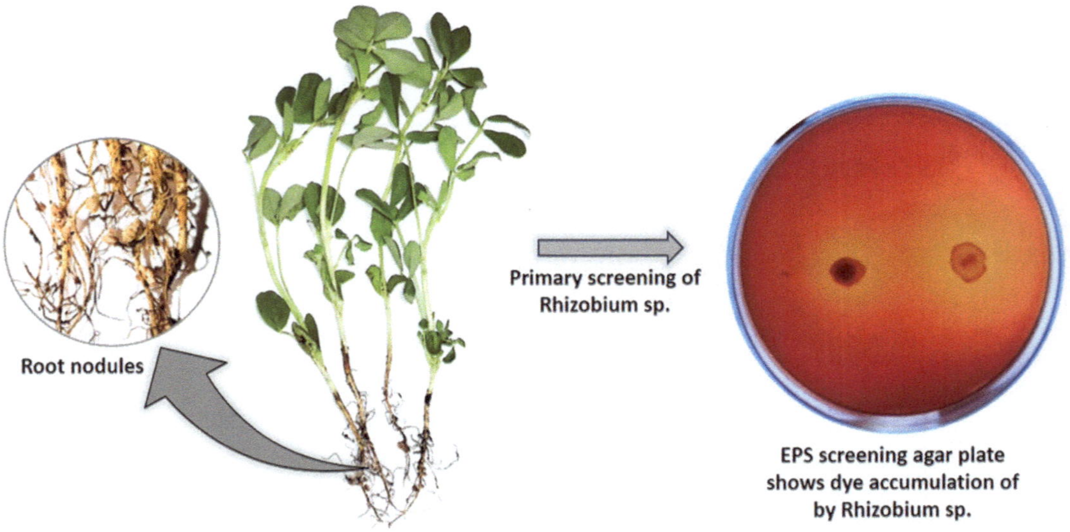

Fig. 1 Zone of dye accumulation surrounding EPS-producing *Rhizobium* on glutamate-containing media with neutral red

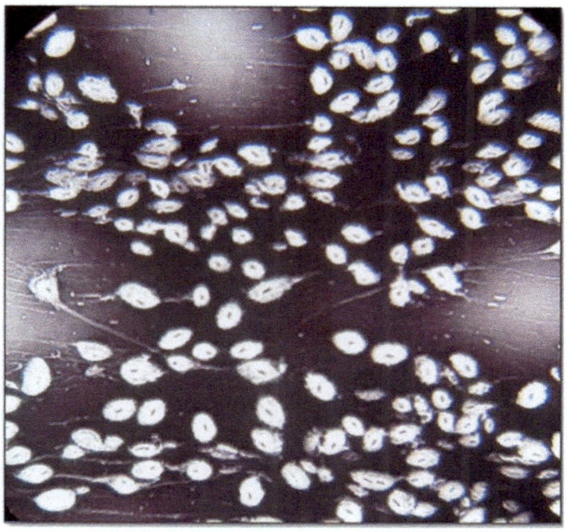

Fig. 2 Demonstration of EPS-producing efficiency (EPS layer surrounding the producing bacterial cell by negative staining)

3. Using a clean, grease-free slide held at a 30° angle, spread a drop of the bacterial suspension forward direction and make a thin film.
4. Allow the stained film to air-dry and observe under the oil immersion lens of a compound microscope. Observe the bacterial cells surrounded by the EPS layer against a dark background. The area of EPS surrounding the cell in the negative stain provides a primary indication of EPS production (Fig. 2).
5. Colonies testing positive for EPS production by negative staining are used for further EPS production in Burk's nitrogen-free liquid medium (*see* **Note 4**). Incubate the culture broth of EPS-producing bacteria for 24–48 h at 30 °C and 120 rpm.
6. Extract and purify the EPS by isopropanol precipitation at a 1: 2 ratio (fermented medium: solvent). Oven-dry the purified EPS at 50 °C for 8 h.
7. Quantify the EPS on a dry weight basis (g/L) by measuring weights on microanalytical balance.

3.4 Isolation and Screening of Cellulose-Producing Bacterial Strain

1. Cultivate the test culture on HS agar medium (*see* **Note 5**) plates, and then inoculate in 100 mL HS liquid media in 250-mL Erlenmeyer flasks. The culture flasks showing bead formation are serially diluted in sterile phosphate buffer.
2. Spread a 0.1 mL aliquot of the 10^{-5} dilution onto HS agar medium supplemented with Congo red. Incubate the plates at 30 °C for 24 h and observe for characteristic bright-red colonies, indicative of cellulose producers. This screening can also be performed in liquid HS medium supplemented with Congo red dye. The specific binding of Congo red to cellulose results

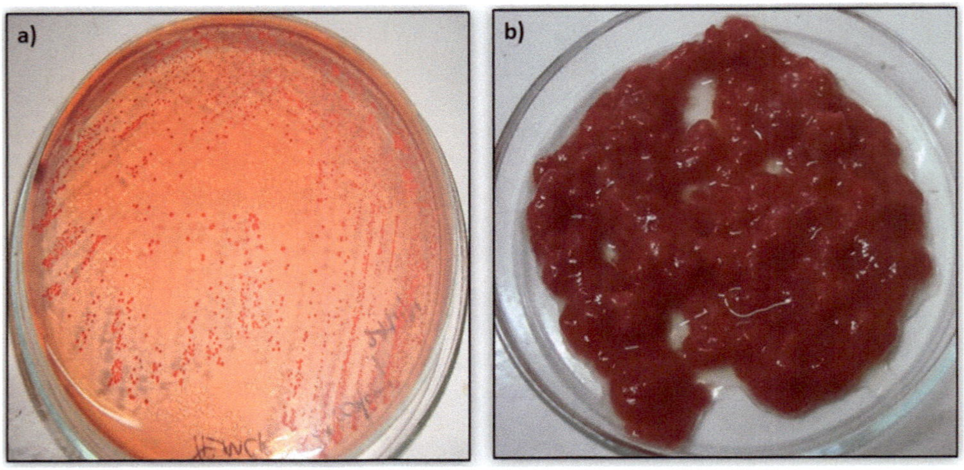

Fig. 3 (**a**) Colonies of cellulose-producing bacteria on screening RMG medium with Congo red. (**b**) Extracted wet red-colored bacterial cellulose produced in HS medium supplemented with Congo red

in the characteristic red coloration of bacterial cellulose (Fig. 3).

3. After 5 days of incubation in a liquid HS medium, separate the produced bacterial cellulose (BC) beads by filtration and rinse them with distilled water to remove residual media. Immediately boil the beads at 90 °C in 0.1 M NaOH solution for 30 min to inactivate any attached bacterial cells.
4. After boiling, purify the BC beads by extensively washing them with distilled water at room temperature until the pH of the wash water becomes neutral. Oven-dry the purified cellulose at 50 °C for 8 h.
5. Quantify the bacterial cellulose on a dry weight basis (g/L) using microanalytical balance.

4 Notes

1. *EPS screening medium 1:* Nutrient agar supplemented with 5.0 g/L monosodium glutamate and either 0.006% (w/v) neutral red or 0.04 g/L methylene blue.
2. *EPS screening medium 2:* Low-strength nutrient agar containing 5.0 g/L monosodium glutamate and either 0.006% (w/v) neutral red or 0.04 g/L methylene blue.
3. *Yeast extract mannitol agar (YEMA) (g/L):* Yeast extract 1.0 g, mannitol 10.0 g, dipotassium phosphate 0.5 g, magnesium sulfate 0.2 g, sodium chloride 0.1 g.

Adjust the final pH to 7.0 ± 0.2 at 25 °C.

Add Congo red (25 mg/L), mix well, and add agar–agar (30.0 g/L). Boil with shaking to dissolve completely.

4. *Nitrogen-free Burk's medium (g/L):* Magnesium sulfate 0.2 g, dipotassium hydrogen phosphate 0.8 g, potassium dihydrogen phosphate 0.2 g, calcium sulfate 0.13 g, iron (III) chloride 0.00145 g, sodium molybdate 0.000253 g, sucrose 20.0 g.
5. *HS medium (g/L):* Glucose 20.0 g, peptone 5.0 g, yeast extract 5.0 g; disodium phosphate (anhydrous) 2.7 g, and citric acid (monohydrate) 1.15 g. Adjust the pH to 5.0 using 0.1 N HCI or 0.1 N NaOH. Add Congo red (0.05 g) after pH adjustment.
6. *For screening of EPS-producing fungal endosymbionts:* Use potato dextrose agar (PDA) supplemented with an additional 1% dextrose. The rest of the screening procedure remains the same as for bacterial endosymbionts.

Acknowledgement

The corresponding author **Prof. (Dr.) Satish V Patil** kindly acknowledges financial support from RGSTC, Government of Maharashtra, File No. KBCNMU/RGSTC/Sanction Order/ 155, Date: 25/07/2023.

References

1. Deka D, Sonowal S, Chikkaputtaiah C, Velmurugan N (2020) Symbiotic associations: key factors that determine physiology and lipid accumulation in oleaginous microorganisms. Front Microbiol 11:555312
2. Stadler M, Schulz B (2009) High energy biofuel from endophytic fungi? Trends Plant Sci 4: 353–355
3. Orlandelli RC, Vasconcelos AFD, Azevedo JL, da Silva MDLC, Pamphile JA (2016) Screening of endophytic sources of exopolysaccharides: preliminary characterization of crude exopolysaccharide produced by submerged culture of Diaporthe sp. JF766998 under different cultivation time. Biochim Open 2:33–40
4. Breedveld MW, Cremers HC, Batley M, Posthumus MA, Zevenhuizen LP, Wijffelman CA et al (1993) Polysaccharide synthesis in relation to nodulation behavior of *Rhizobium leguminosarum*. J Bacteriol 175:750–757
5. Breierová E, Hromádková Z, Stratilová E, Sasinková V, Ebringerová A (2005) Effect of salt stress on the production and properties of extracellular polysaccharides produced by *Cryptococcus laurentii*. Z Naturforsch C 60: 444–450
6. Mohite BV, Koli SH, Patil SV (2018) Heavy metal stress and its consequences on exopolysaccharide (EPS)-producing *Pantoea agglomerans*. Appl Biochem Biotechnol 186:199–216
7. Castellani LG, Luchetti A, Nilsson JF, Pérez-Giménez J, Wegener C, Schlüter A et al (2021) Exopolysaccharide characterization of *rhizobium favelukesii* LPU83 and its role in the symbiosis with alfalfa. Front Plant Sci 12: 642576
8. Fraysse N, Couderc F, Poinsot V (2003) Surface polysaccharide involvement in establishing the rhizobium–legume symbiosis. Eur J Biochem 270:1365–1380
9. Costa OY, Raaijmakers JM, Kuramae EE (2018) Microbial extracellular polymeric substances: ecological function and impact on soil aggregation. Front Microbiol 9:1636
10. Hawkins JP, Oresnik IJ (2022) The rhizobium-legume symbiosis: co-opting successful stress management. Front Plant Sci 12:796045

11. Ghosh PK, Maiti TK (2016) Structure of extracellular polysaccharides (EPS) produced by rhizobia and their functions in legume–bacteria symbiosis: a review. ALS 10:136–143
12. Skorupska A, Janczarek M, Marczak M, Mazur A, Król J (2006) Rhizobial exopolysaccharides: genetic control and symbiotic functions. Microb Cell Factories 5:1–19
13. D'Haeze W, Glushka J, De Rycke R, Holsters M, Carlson RW (2004) Structural characterization of extracellular polysaccharides of *Azorhizobium caulinodans* and importance for nodule initiation on *Sesbania rostrata*. Mol Microbiol 52:485–500
14. Pellock BJ, Cheng HP, Walker GC (2000) Alfalfa root nodule invasion efficiency is dependent on *Sinorhizobium meliloti* polysaccharides. J Bacteriol 182:4310–4318

Chapter 19

Screening of Endosymbionts for Biocontrol Properties

Neelam Thakur, Simranjeet Kaur, and Samiksha Jhamta

Abstract

Endosymbiotic bacteria *Xenorhabdus* and *Photorhabdus* spp. extracted from the entomopathogenic nematodes can be a promising tool for pest management strategies. The endosymbionts need to be screened for their biocontrol activities after their isolation. This chapter incorporates the methods for the isolation of entomopathogenic nematodes, their endosymbiont bacteria, and further their screening procedure. These screening methods are based on the various secondary metabolites secreted by the endosymbionts. Secondary metabolites such as ammonia, siderophores, hydrogen cyanide (HCN), and hydrolytic enzymes (amylase, chitinase, and proteases) attribute to the antimicrobial properties. Endosymbionts screened for biocontrol activities can play a pivotal role in the direction of sustainable agriculture and biological control.

Key words Biocontrol, Chitinase, Entomopathogenic nematodes, Endosymbionts, HCN, Hydrolytic enzymes, Screening, Siderophores

1 Introduction

The search for alternatives to chemical pesticides amid the splurging demand for food, pest resistance, good soil, and human health has led us to the beneficial nematodes [1]. These entomopathogenic nematodes (EPNs) belonging to the genera *Steinernema* and *Heterorhabditis* especially with their endosymbiont bacteria *Xenorhabdus* and *Photorhabdus* spp., respectively, are the future of pest management strategies [2]. These bacterial endosymbionts aid in a number of benefits to the associated plant including defense against natural enemies [3]. In addition to insects, these bacterial symbionts also protect the plant from parasitic nematodes, viruses, pathogenic fungi, parasitoids, and other predators [4]. These beneficial endosymbionts secrete a number of secondary metabolites which enhance their biocontrol properties [5, 6], and detection of these can determine the biocontrol potential of any isolated bacteria. The endosymbionts can be screened for various attributes such as ammonia production, siderophore production, hydrogen cyanide (HCN) production, and production of hydrolytic enzymes

N. Amaresan and Sapna Chandwani (eds.), *Endosymbionts*, Springer Protocols Handbooks,
https://doi.org/10.1007/978-1-0716-4771-4_19,

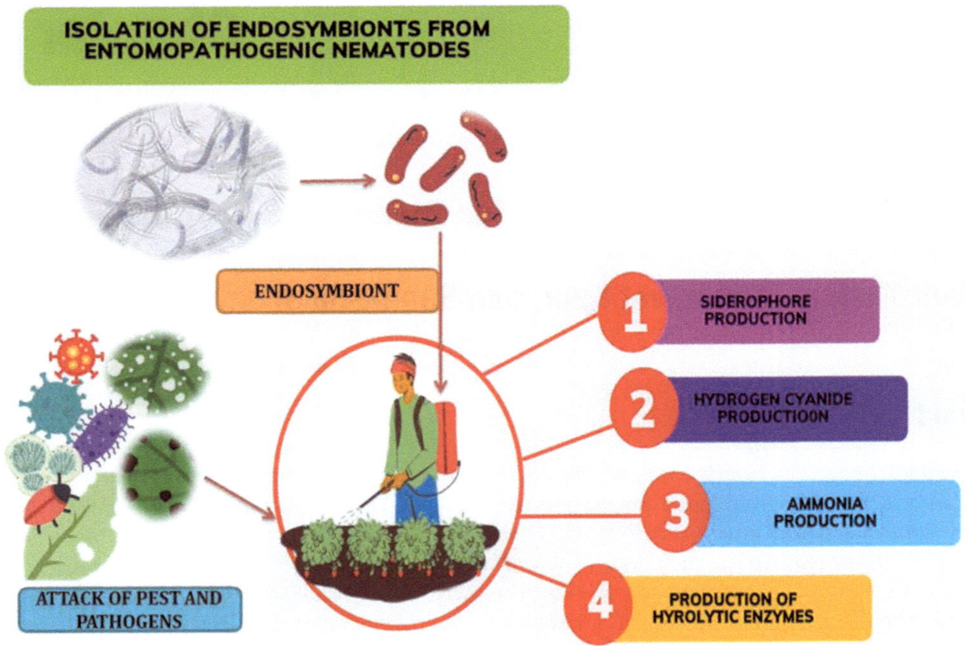

Fig. 1 Endosymbionts-Screening for biocontrol properties

(chitinase amylase and proteases) to ensure their role in induced defense mechanism against insects.

The siderophore production by bacteria is one of the important mechanisms of biological control of pathogens as it facilitates the competition for iron with other soil-borne pathogens. Also, it is a source of iron for plants under iron stress which further prevents root proliferation by phytopathogens [7]. Bacterial HCN is produced through a metabolic pathway of cyanogenesis and is linked with the suppression of growth and death of other living organisms [8]. Ammonia production is also correlated with the impaired growth of several plant fungal pathogens [9].

Microorganisms produce numerous hydrolytic enzymes including chitinase, amylase, and protease. The chitinase tends to cause cell wall weakening and degradation of numerous pests and pathogens. Thus, it is known to exhibit insecticidal, antibacterial, antifungal, and nematicidal properties [10]. Amylase production in bacteria tends to plant growth promotion and prevention of plant diseases [11]. Protease production of bacteria assists in its antimicrobial activity against pathogens such as bacteria, fungi, viruses, and nematodes [12].

This chapter summarizes the isolation of endosymbiotic bacteria from EPNs and their screening for biocontrol properties. The screening method utilizes a number of tests discussed along with the concerned protocol (Fig. 1).

2 Materials

The screening process of endosymbiont can only be initiated after the isolation of nematodes. The isolation of entomopathogenic nematodes from the soil is done using the insect soil baiting technique. After the initial isolation of EPNs, their endosymbiont bacterium is extracted from the nematodes. Isolation of entomopathogenic microbes is carried out using growth media and nutrient agar added with 0.004% (w/v) triphenyltetrazolium chloride and 0.025% (w/v) bromothymol blue [13]. The isolated bacteria are further subjected to qualitative and quantitative screening using attributes such as the production of siderophore ammonia, HCN, and hydrolytic enzymes (chitinase, amylase, proteases). The positive screened microbial isolates are further identified and stored for biocontrol activities. The following Table 1 shows the materials required for the isolation and screening of endosymbionts.

3 Methods

3.1 Isolation of Entomopathogenic Nematodes Using Insect Soil Baiting Technique

1. Collect the soil samples from different agriculture fields, orchards, and most preferably the undisturbed sites (*see* **Note 1**). Randomly collect the samples at a depth of 15–30 cm near the rhizospheric soil near the vegetation [14].
2. Measure about 200 g of soil and keep it in plastic jars (*see* **Note 2**). The last instar of wax moth (*Galleria mellonella*) or rice moth (*Corcyra cephalonica*) larvae are the most used insect bait for EPN isolation.
3. Add approximately five to ten larvae of bait insect into each jar and pour the collected soil on it carefully (*see* **Note 1**). Later, mark the jars with the sample name, number, date and type of bait, etc. Extract the infective juveniles (IJs) by using the modified white trap method [14, 15].
4. For setting up the white trap, two different-sized Petri plates and one Whatman filter paper are used. The smaller Petri plate (60 mm) is lined with a circular Whatman filter paper no. 1 and is placed inside a larger Petri plate (100 × 15 mm) (*see* **Note 3**).
5. The larger Petri plate is filled with about 200 mL of sterile distilled water. Cover the larger Petri plate with the lid and label the plate with the sample name, date of setting the trap, etc.
6. The white traps are then kept at room temperature until the emergence of IJs. The process of emergence can take about 10–25 days depending upon the type of isolate/species.
7. Harvesting of IJs is done by removing the water from the large Petri plates into a beaker. The nematodes are allowed to settle

Table 1
Material required for the isolation and screening of endosymbionts

Attributes	Material required
Isolation of EPNs	Insect bait, polythene bags for soil collection, plastic jars, marker, tags, petri plates (100 × 15 mm and 60 mm), Whatman filter paper no. 1, distilled water, beaker, dropper
Composition of NBTA	1 L nutrient agar, 0.025 g bromothymol blue (BTB), 0.04 g triphenyltetrazolium chloride (TTC)
Siderophore screening	6 M HCl Blue dye: 0.06 g CAS, 0.0027 g $FeCl_3$-$6H_2O$, 10 mM HCl, 0.073 g of HDTMA Minimal media 9 (MM9) salt solution stock: 15 g KH_2PO_4, 50 g NH_4Cl, and 25 g NaCl dissolved in 500 mL ddH_2O NaOH stock: 25 g of NaOH dissolved in 150 mL ddH_2O; pH 12 20% glucose stock: 20 g glucose dissolved in 100 mL ddH_2O Casamino acid solution: 3 g casamino acid dissolved in 27 mL ddH_2O, 3% 8-hydroxyquinoline in chloroform should be used to remove any trace iron CAS agar preparation: 100 mL MM9 salt solution dissolved in 750 mL ddH_2O, 15 g Bacto agar, 32.24 g piperazine-N,N′-bis (2-ethanesulfonic acid) PIPES, 30 mL casamino acid solution
HCN screening	King's B media with 4.5 g glycine in 1 L ddH_2O, 2% sodium carbonate, 0.5% picric acid
Ammonia screening	1.5% w/v peptone water, 5% Nessler reagent
Chitinase screening	Colloidal chitin: Chitin powder 40 g, concentrated HCl 600 mL, distilled water Chitin media: 0.1% KH_2PO_4, 0.1% $MgSO_4$, 0.2% K_2HPO_4, 0.05% P, 1% $NaNO_3$, 0.05% N, and 2% agar
Amylase screening	5.0 g peptone, 1.5 g yeast extract, 2.0 g starch, 1.5 g meat extract, 5.0 g NaCl, 15 g agar,1 L ddH_2O
Protease screening	10 g glucose, 5 g yeast extract, 5 g skim milk, 2 g KH_2PO_4, 10 g Na_2CO_3, 1 L ddH_2O

down for about 10–20 min. After this, the water is carefully poured off, as the nematodes have settled at the bottom of the beaker. This step is reiterated two to three times till the water becomes clear and clean.

8. After recovery, the nematodes are stored in a conical flask in the form of suspension for later use.

3.2 Isolation of Endosymbiont from EPNs

1. The isolation of endosymbiont from EPNs requires the preparation of NBTA media. The composition is given in Table 1. Prepare nutrient agar with BTB and autoclave this mixture at 121 °C for 15 min.
2. Later, add the TTC, just before pouring the media into Petri dishes using a milipore filter (*see* **Note 4**). Swirl the media and then dispense it into sterile Petri dishes. Suspend surface-

sterilized IJs into nutrient agar and homogenate the suspension in a sterile tissue homogenizer (*see* **Note 5**).

3. Transfer the suspension into a test tube and allow the bacteria to multiply on a shaker for 24–48 h. The isolated bacteria are further streaked on NBTA to obtain the formation of individual colonies.

The streaked Petri plates are incubated at 25–28 °C for 3–5 days. The isolated bacteria are further preserved in broth and glycerol for further use [14].

3.3 Screening for Siderophore Production

Preparation of CAS media: The preparation requires the making of different stock solutions and blue dye. The procedure is as follows:

- *Blue dye:* For the preparation of blue dye, we need to make three solutions: A, B, and C. Solution A: In a beaker, dissolve 0.06 g of CAS in 50 mL of ddH_2O. Solution B: Dissolve 0.0027 g of $FeCl_3$-6H_2O in 10 mL of 10 mM HCl. Solution C: Dissolve 0.073 g of HDTMA in 40 mL of ddH_2O. Now, after making the three solutions, mix Solution A with 9 mL of Solution B. Then, mix with Solution C. The resulting mixture should be of blue color. Autoclave the dye and store it in a reagent bottle (*see* **Note 6**).
- *Mixture solution:* The mixture solution contains different stock solutions.
  1. *Minimal media 9 (MM9) salt solution stock:* Dissolve 15 g KH_2PO_4, 25 g NaCl, and 50 g NH_4Cl in 500 mL of ddH_2O.
  2. *20% glucose stock:* Dissolve 20 g glucose in 100 mL of ddH_2O.
  3. *NaOH stock:* Dissolve 25 g of NaOH in 150 mL ddH_2O; pH 12.
  4. *Casamino acid solution:* Dissolve 3 g of casamino acid in 27 mL of ddH_2O. Extract the solution with 3% 8-hydroxyquinoline and filter sterilize it.
- *CAS agar preparation:* First add 100 mL of MM9 salt solution to 750 mL of ddH_2O. Then, dissolve 32.24 g piperazine-N, N′-bis (2-ethanesulfonic acid) PIPES in it. After this, add 15 g of agar powder. Autoclave the media and, after cooling it at 50 °C, add 30 mL of sterile casamino acid solution and 10 mL of sterile 20% glucose. In the last, slowly add 100 mL of blue dye solution along the glass wall with enough agitation to mix thoroughly and pour in the Petri dishes (*see* **Note 7**).
- Siderophore production is tested using this prepared chrome azurol S (CAS) agar plates. The isolated bacterial culture is

spot-inoculated on the CAS agar plates and incubated at 37 °C for a period of 12 days. The appearance of orange halos around the colonies on the blue-colored agar indicates siderophore production [16].

3.4 Screening for Hydrogen Cyanide Production

1. King's B medium with 4.5 g/L glycine is prepared and filled in the test tubes. The isolated bacteria are inoculated in the test tubes, and filter paper soaked in 0.5% picric acid and 2% sodium carbonate solution is positioned at the test tube's top and sealed using parafilm.
2. The inoculated test tubes are incubated at 37 °C for a duration of 4 days. The development of yellow to red color on the filter paper indicates the isolate to be positive for HCN production [17].

3.5 Screening for Ammonia Production

1. The ammonia production can be determined using peptone water.
2. The isolated bacteria are inoculated in peptone water (1.5% w/v) and further incubated at 30 °C for a time period of 2 days on a shaker at a speed of 150 rpm.
3. After the stated duration, 5% Nessler's reagent is added to the cell-free supernatant of fermented broth.
4. The change in color from pale to deep yellow indicates the production of ammonia [18].

3.6 Screening for Hydrolytic Enzymes

1. *Chitinase production.*
   - In the first step, the colloidal chitin needs to be prepared. The loose colloidal chitin obtained is further used as a substrate.
   - The screening is performed with bacterial isolates on the colloidal chitin agar medium which is incubated at 37 °C.
   - Bacterial isolates are selected on the basis of a larger hydrolysis zone after 96 h of incubation and further screened for maximum enzyme production in nutrient broth media.
   - The cultures are centrifuged at 10,000 rpm for 15 min at 4 °C, and the crude enzyme extracted is used for chitinase assay.
   - The chitinase activity is assessed by measuring reducing sugar released from colloidal chitin.
   - The crude enzyme (150 μL) is added to the mixture consisting of 300 μL of 0.1% colloidal chitin and 150 μL of 0.1 M phosphate buffer, pH 7.0. After incubation at 55 °C for 10 min, the reaction mixture is subjected to a refrigerated centrifugation at 10,000 rpm for 5 min.

- The resulting supernatant (200 μL) is added with 500 mL of distilled water and 1000 mL of Schales reagent and then boiled for 10 min. After cooling, the absorbance of the mixture is measured at 420 nm [19].

2. *Amylase production.*
   - The production of amylase is evaluated in a starch medium (Table 1). A single colony from the isolated bacteria is spotted at the center of a Petri dish containing starch medium. Plates are incubated at 30 °C for 24 h.
   - After the incubation, an iodine solution is poured onto the plate until the colony is submerged. It is then left for 1 min. If a clear halo is observed around the colony, the isolate is considered positive for amylase production [11].
3. *Protease production.*
   - The production of protease is evaluated on skim milk agar medium (Table 1). A single colony from the isolated bacteria is spotted at the center of a Petri dish containing skim milk agar medium.
   - Plates are incubated at 30 °C for 24 h. The appearance of clear zones on skim milk agar plates is the first sign of proteolytic activity [12].

4 Notes

1. The soil samples should be collected from undisturbed sites for a better probability of locating the presence of EPNs [14].
2. Plastic containers or jars should be covered with a dark cloth and kept at room temperature during the insect baiting technique [14].
3. The cadavers recovered from soil after insect baiting are placed on the smaller Petri plate in a way that they do not touch each other, to avoid contamination [4].
4. TTC should be added to the autoclaved cooled media with the help of a millipore filter.
5. TTC will degrade if it is added when agar is too hot.
6. A quality control test is needed for CAS media as it is quite sensitive to fluctuations in $FeCl_3$ concentrations or pH.
7. To remove any trace elements while preparing for CAS media, all the glassware should be cleaned with 6 M HCl and then rinsed with double distilled water [16].

Acknowledgement

The essential laboratory facilities were provided by the vice chancellor and pro vice chancellor of Eternal University, Baru Sahib, for which the authors are grateful.

References

1. Thakur N, Kaur S, Tomar P et al (2020) Microbial biopesticides: current status and advancement for sustainable agriculture and environment. In: Asghar R, Nath YA, Neelam Y (eds) New and future developments in microbial biotechnology and bioengineering, vol 1. Elsevier, Amsterdam, pp 243–282
2. Gümüşsoy A, Yüksel E, Özer G, İmren M, Canhilal R, Amer M et al (2022) Identification and biocontrol potential of entomopathogenic nematodes and their endosymbiotic bacteria in apple orchards against the codling moth, *Cydia pomonella* (L.)(Lepidoptera: Tortricidae). Insects 13:1085
3. Kour D, Negi R, Khan SS, Kumar S, Kaur S, Kaur T et al (2023) Microbes mediated induced systemic response in plants: a review. Plant Stress 11:100334
4. White JA, Richards NK, Laugraud A, Saeed A, Curry MM, McNeill MR (2015) Endosymbiotic candidates for parasitoid defense in exotic and native New Zealand weevils. Microbial Ecol 70:274–286
5. Tomar P, Thakur N, Yadav AN (2022) Endosymbiotic microbes from entomopathogenic nematode (EPNs) and their applications as biocontrol agents for agro-environmental sustainability. Egypt J Biol Pest Control 32:80
6. Thakur N, Tomar P, Kaur S, Jhamta S, Thakur R, Yadav AN (2021) Entomopathogenic soil microbes for sustainable crop protection. In: Yadav AN (ed) Soil microbiomes for sustainable agriculture, vol 1. Springer, Cham, pp 529–571
7. Sasirekha B, Srividya S (2016) Siderophore production by *Pseudomonas aeruginosa* FP6, a biocontrol strain for *Rhizoctonia solani* and *Colletotrichum gloeosporioides* causing diseases in chilli. Agric Nat Resour 50:250–256
8. Sehrawat A, Sindhu SS, Glick BR (2022) Hydrogen cyanide production by soil bacteria: biological control of pests and promotion of plant growth in sustainable agriculture. Pedosphere 32:15–38
9. Lee JH, Anderson AJ, Kim YC (2022) Root-associated bacteria are biocontrol agents for multiple plant pests. Microorganisms 10:1053
10. Veliz EA, Martínez-Hidalgo P, Hirsch AM (2017) Chitinase-producing bacteria and their role in biocontrol. AIMS Microbiol 3:689
11. Rafanomezantsoa P, Gharbi S, Karkachi N, Kihal M (2023) Optimization of amylase production by the biological control agent *Bacillus halotolerans* RFP74 using response surface methodology. J Genet Eng and Biotechnol 21:63
12. Afrin S, Tamanna T, Shahajadi UF, Bhowmik B, Jui AH, Miah MAS et al (2024) Characterization of protease-producing bacteria from garden soil and antagonistic activity against pathogenic bacteria. The Microbe 4: 100123
13. Kaya HK, Stock SP (1997) Techniques in insect nematology. In: Lacey LA (ed) Manual of techniques in insect pathology. Elsevier, San Diego, pp 281–324
14. Orozco RA, Lee M-M, Stock SP (2014) Soil sampling and isolation of entomopathogenic nematodes (Steinernematidae, Heterorhabditidae). J Vis Exp 89:e5208
15. White G (1927) A method for obtaining infective nematode larvae from cultures. Science 66: 302–303
16. Schwyn B, Neilands J (1987) Universal chemical assay for the detection and determination of siderophores. Anal Biochem 160:47–56
17. Bakker AW, Schippers B (1987) Microbial cyanide production in the rhizosphere in relation to potato yield reduction and *Pseudomonas* spp-mediated plant growth-stimulation. Soil Biol Biochem 19:451–457
18. Ahmad F, Ahmad I, Khan M (2008) Screening of free-living rhizospheric bacteria for their multiple plant growth promoting activities. Microbiol Res 163:173–181
19. Kuddus M, Ahmad I (2013) Isolation of novel chitinolytic bacteria and production optimization of extracellular chitinase. J Genet Eng Biotechnol 11:39–46

Chapter 20

Screening of Endosymbionts for Virucidal Activity

Hetvi Naik, Sapna Chandwani, and Natarajan Amaresan

Abstract

Endosymbiotic microorganisms from various host organisms have emerged as promising sources of novel antiviral compounds. This chapter describes methods for screening endosymbionts from diverse hosts, including insects, plants, marine invertebrates, and mammals, for virucidal activity against model viruses. The protocol includes techniques for preparing endosymbiont extracts, conducting plaque reduction assays, and performing preliminary characterization of active compounds.

Key words Endosymbionts, Virucidal activity, Plaque assay

1 Introduction

Endosymbiotic microorganisms living within various host organisms have emerged as a promising source of novel bioactive compounds, including those with potential antiviral properties [1]. These endosymbionts, found in hosts ranging from insects and plants to marine invertebrates and mammals, have co-evolved with their hosts over millions of years and often develop specialized metabolic capabilities [2, 3]. The exploration of endosymbionts for antimicrobial compounds has gained traction in recent years. Studies have demonstrated that endosymbionts from various sources produce compounds with antiviral activity. For example, fungal endophytes isolated from plants have shown activity against influenza viruses [4], while bacteria associated with marine sponges have produced compounds effective against herpes simplex virus [5].

Insect endosymbionts have received considerable attention due to their diversity and the evolutionary pressures that may have led to the development of antimicrobial defenses [6]. For instance, *Wolbachia*, a common insect endosymbiont, has been shown to confer antiviral protection to its host [7]. Mammalian gut microbiota, while not traditionally considered endosymbionts, also play a crucial role in antiviral defense. Recent studies have shown that

N. Amaresan and Sapna Chandwani (eds.), *Endosymbionts*, Springer Protocols Handbooks,
https://doi.org/10.1007/978-1-0716-4771-4_20,

commensal bacteria can enhance antiviral immunity and even produce compounds with direct antiviral activity [8]. Virucidal compounds are particularly valuable as they can directly inactivate viral particles, potentially preventing infection [9]. While most antiviral drugs target specific stages of the viral life cycle within host cells, virucidal agents can act as broad-spectrum antivirals by disrupting viral structural integrity [10].

2 Materials

Prepare all solutions using ultrapure water (to prepare ultrapure water, purify deionized water, to attain a sensitivity of 18 MΩ^{-cm} at 25 °C) and analytical-grade reagents. Prepare and store all reagents at room temperature (unless indicated otherwise).

2.1 Endosymbiont Extract Preparation

1. Tryptic soy broth

Peptone	17.0 g/L
Enzymatic digest of soy	3.0 g/L
Sodium chloride	5.0 g/L
Dipotassium phosphate	2.5 g/L
Dextrose	2.5 g/L
Final pH (at 25 °C)	7.3 ± 0.2

2. *Marine broth*

Peptone	5 g/L
Sodium chloride	19.45 g/L
Magnesium chloride	8.8 g/L
Sodium sulfate	3.24 g/L
Calcium chloride	1.8 g/L
Potassium chloride	0.55 g/L
Sodium bicarbonate	0.16 g/L
Ferric citrate	0.1 g/L
Final pH (at 25 °C)	7.6 ± 0.2

3. *Potato dextrose broth*

Potato infusion	200.0 g/L

(continued)

Dextrose	20.0 g/L
Yeast extract	4.0 g/L
Final pH (at 25 °C)	5.6 ± 0.2

4. Shaking incubator.
5. Centrifuge.
6. 0.22-μm filters.
7. Organic solvents (ethyl acetate, methanol).

2.2 Primary Screening (Cell Viability Assay)

1. 96-well cell culture plates.
2. Vero cells or other appropriate cell lines.
3. Cell culture medium (e.g., DMEM with 10% FBS).
4. Model virus (e.g., Herpes simplex virus 1).
5. Endosymbiont extracts.
6. MTT (3-(4,5-dimethylthiazol-2-yl)-2,5-diphenyltetrazolium bromide) or similar cell viability reagent.
7. Microplate reader.

2.3 Secondary Screening (Plaque Reduction Assay)

1. 6-well cell culture plates.
2. Vero cells.
3. Model virus.
4. Methyl cellulose overlay media.
5. Crystal violet stain.

3 Methods

3.1 Preparation of Endosymbiont Extract

1. Culture endosymbionts isolated from various hosts in appropriate media for 24–72 h (*see* **Note 1**).
2. Centrifuge cultures at 4000×*g* for 15 minutes to pellet the cells.
3. For bacterial and fungal cultures, filter sterilize supernatants through a 0.22-μm filter.
4. For fungal cultures and slow-growing bacteria, extract metabolites from both cells and supernatant using ethyl acetate or methanol.
5. Concentrate extracts using a rotary evaporator and resuspend in a small volume of appropriate solvent (DMSO).

3.2 Primary Screening: Cell Viability Assay

1. Seed Vero cells in 96-well plates (10,000 cells/well) and incubate overnight.

2. Prepare virus stock dilutions to yield a multiplicity of infection (MOI) of 0.1.
3. In separate plates, mix equal volumes (50 μL) of virus dilution and endosymbiont extract. Include solvent-only and virus-only controls (*see* **Note 2**).
4. Incubate virus-extract mixtures at 37 °C for 1 h.
5. Remove media from cell plates and add 100 μL of virus-extract mixture to each well.
6. Incubate plates at 37 °C for 48 h.
7. Add MTT reagent and incubate for 4 h.
8. Measure the absorbance at 570 nm using a microplate reader.
9. Calculate percent viability compared to uninfected control.
10. Cytotoxicity of extracts on uninfected cells was assessed in parallel assays (*see* **Note 3**).

3.3 Plaque Reduction Assay

1. Select extracts showing >50% protection in the primary screening.
2. Grow Vero cells to confluency in 6-well plates.
3. Prepare virus stock dilutions to yield 100–200 plaque-forming units (PFU) per well.
4. Mix equal volumes of diluted viral extract and endosymbiont extract. Include solvent-only as control (*see* **Note 4**).
5. Incubate virus-extract mixtures at 37 °C for 1 h.
6. Infect cell monolayers with virus-extract mixtures for 1 h at 37 °C, remove inoculum and overlay with methyl cellulose media again, and incubate for 48–72 h at 37 °C.
7. Fix and stain cells with crystal violet.
8. Count plaques and calculate percent reduction compared to solvent-only control (*see* **Notes 5, 6**).

4 Notes

1. Use appropriate biosafety precautions when handling microbes and viruses.
2. Include positive (known antiviral) and negative (media only) controls in all assays.
3. Assess the cytotoxicity of extracts on uninfected cells in parallel with antiviral assays.
4. Confirm virucidal activity by pre-treating the virus before cell infection.

5. Verify the reproducibility of active extracts with multiple independent cultures.
6. Consider testing active extracts against a panel of different viruses to determine specificity.
7. For marine-derived samples, test for salt tolerance of the virus and host cells used in the assay.

References

1. Stierle AA, Stierle DB (2015) Bioactive secondary metabolites produced by the fungal endophytes of conifers. Nat Prod Commun 10:1671–1682
2. Moran NA, McCutcheon JP, Nakabachi A (2008) Genomics and evolution of heritable bacterial symbionts. Annu Rev Genet 42:165–190
3. Flórez LV, Biedermann PHW, Engl T et al (2015) Defensive symbioses of animals with prokaryotic and eukaryotic microorganisms. Nat Prod Rep 32:904–936
4. Dey P, Kundu A, Kumar A et al (2020) Analysis of alkaloids (indole alkaloids, isoquinoline alkaloids, tropane alkaloids). In: Recent advances in natural products analysis. Elsevier, Amsterdam, pp 505–567
5. Chevrette MG, Carlson CM, Ortega HE et al (2019) The antimicrobial potential of Streptomyces from insect microbiomes. Nat Commun 10:516
6. Kaltenpoth M, Engl T (2014) Defensive microbial symbionts in Hymenoptera. Funct Ecol 28:315–327
7. Hedges LM, Brownlie JC, O'Neill SL et al (2008) *Wolbachia* and virus protection in insects. Science 322:702–702
8. Pfeiffer JK, Virgin HW (2016) Transkingdom control of viral infection and immunity in the mammalian intestine. Science 351:aad5872
9. Blaising J, Polyak SJ, Pécheur E-I (2014) Arbidol as a broad-spectrum antiviral: an update. Antivir Res 107:84–94
10. Yacoby I, Bar H, Benhar I (2007) Targeted drug-carrying bacteriophages as antibacterial Nanomedicines. Antimicrob Agents Chemother 51:2156–2163

Chapter 21

Screening of Endosymbionts for Bactericidal Activity

Lizzy A. Mwamburi

Abstract

Because of the rapid increase in microbial resistance to existing antimicrobial agents, there is a need to explore unique sources of antibiotics from nature. Hence, current scientific endeavors have focused on the bioprospecting of secondary metabolites found among microbes, with the special aim of finding naturally occurring compounds possessing antibacterial qualities that originate from endosymbionts. All plant species, regardless of their type, are believed to harbor endosymbiotic microorganisms. In addition to the benefits, they confer to the host plants, i.e., fungal endosymbionts produce a wide range of natural chemicals for exploitation with possible utilization in the fields of medicine, agriculture, and pharmaceuticals. The procedures described in the current protocols focus on examining the antibacterial properties of fungal endosymbionts isolated in higher plants.

Key words Antibacterial activity, Endosymbiont fungi, Agar plug diffusion, Disc diffusion method, Agar dilution method

1 Introduction

Endosymbionts are organisms that reside inside the cells of plantae. These microorganisms, usually fungus and/or bacteria, form beneficial connections with their plant host, leading to enhanced host fitness. Endosymbionts can finish part of their life cycle or perhaps the entire life cycle inside their host without exhibiting any symptoms of illness; they are therefore regarded as nonpathogenic.

Fungal endosymbionts belonging to the phyla Ascomycota, Basidiomycota, and Mucoromycota [1] constitute a large component of the plant endophytic community [2, 3]. Endosymbiotic fungi have been recognized for their multiple beneficial roles in plant–microbe interactions including enhancing plant growth and development [4, 5], competitiveness of the host plants against herbivores [6–9], inducing systemic resistance against biotic [10] and abiotic stresses [11–15], and biocontrol activity against pests and pathogens [9, 16–20] in exchange for shelter and nutrition from host plants [21]. They are known to produce secondary

N. Amaresan and Sapna Chandwani (eds.), *Endosymbionts*, Springer Protocols Handbooks,
https://doi.org/10.1007/978-1-0716-4771-4_21,

metabolites [22] which have anticancer, antimicrobial, antiplasmodial, and antiviral activity in the form of alkaloids, macrolides, terpenoids, and peptide derivatives that may be exploited for agricultural and medicinal purposes. The purpose of these protocols is to establish the antibacterial properties of endosymbiont fungus isolated from various plants using the agar disc diffusion, agar plug diffusion, and agar dilution methods.

To investigate the antibacterial properties of endophytes, plants are collected and then identified using morphological methods and taxonomical keys. Plant samples are then dissected into leaves, stems, roots, and flowers. These are disinfected, then cut into 2-mm pieces, cultured on medium, and then incubated at 28 °C for up to 4 weeks [23] in order to isolate the fungal endophytes. The fungal endophytes are identified using morphological and molecular methods. Antibacterial properties of the endophytes are evaluated using various methods as outlined in Subheading 3.

The agar disc diffusion method is a widely used method that involves inoculating agar plates with a standardized inoculum of the test microorganism. The antimicrobial substance is added to the filter paper discs and placed on the agar surface at a specified concentration. The Petri dishes are then incubated in the right environmental conditions. The test microorganisms' germination and growth are usually inhibited by the antimicrobial substance that diffuses into the agar. The diameters of the inhibitory growth zones are then determined.

Although the agar disc diffusion method is not appropriate to determine the minimum inhibitory concentration (MIC), as it is impossible to quantify the amount of the antimicrobial agent diffused into the agar medium, it can be used to test and compare several microorganisms and antimicrobial agents.

The agar plug diffusion method is a popular technique for demonstrating the antagonistic relationship between microorganisms. It entails creating tight streaks on the plate surface to establish an agar culture of the strain of interest on the proper culture medium. Microbial cells release chemicals that permeate the agar media as they proliferate. Following incubation, an aseptic agar cylinder is cut with a sterile cork borer and placed on the agar surface of a different plate that has already been infected with the test microorganism. Diffusion occurs between the chemicals and the agar media. Next, the emergence of the inhibition zone surrounding the agar plug indicates the antibacterial activity of the chemicals released by the microorganisms.

Dilution procedures are suitable for estimating the concentration of the test antimicrobial agent in the agar or broth medium; they are the most suitable for determining minimum inhibitory concentration (MIC) values. The in vitro antibacterial activity can be quantitatively measured using the broth or agar dilution method. The agar dilution method uses serial twofold dilutions by adding the antimicrobial agent to an agar medium in different concentrations. A predetermined microbial inoculum is then inoculated onto the surface of the agar plate. The MIC, which is the lowest concentration of an antimicrobial agent that totally prevents growth under ideal incubation circumstances, is then determined.

2 Materials

2.1 Collection of Plant Material

1. Whole plants at different stages of growth.
2. Sterile polythene bags (for sample collection).
3. Cooler box.
4. Refrigerator (4 °C) for storage.

2.2 Surface Sterilization of Plant Material

1. Running tap water.
2. Ethanol (70%).
3. Sodium hypochlorite (4% active chlorine).
4. Sterile demineralized water.
5. Filter paper.
6. Spatula.
7. Measuring cylinder.
8. 1-liter conical flasks for media preparation.
9. Biomalt agar (Biomalt, 120 g/L BactoTM agar, pH 5.6 with the antibiotics penicillin G Na 60 mg/L, streptomycin sulfate 80 mg/L, and (oxy)-tetracycline-HCl 50 mg/L) for sterility check.
10. pH meter.
11. Sterile Petri dishes.
12. Sterile blades.
13. Sterile plastic bags (for storage).
14. 500-mL Erlenmeyer flask.
15. Forceps.
16. Alcohol lamp.

2.3 Endophytic Fungal Isolation and Cultivation

1. Aqueous agar (distilled water and 1.5% agar–agar).
2. Surface sterilized stem, root, and leaf sections.
3. Measuring cylinder.
4. 1-liter conical flasks for media preparation.
5. Spatula.
6. Streptomycin (3 mg/100 mL).

7. Sterile Petri dishes.
8. Incubator.
9. Mycological medium, (malt extract agar (MEA), potato dextrose agar (PDA), or Sabouraud agar (SA)).
10. Chloramphenicol (0.2 g/L).
11. pH meter.
12. Biomalt agar.
13. Sterile slant tubes.
14. Sterile scissors.
15. For a description of morphological characteristics refer to fungal identification keys.

2.4 Endosymbiont Fungal Cultivation and Extraction

1. Sterile cork borer.
2. 250-mL Erlenmeyer flasks.
3. Potato dextrose broth (PDB).
4. Rotary shaker at 120 rpm.
5. Sterile 15-mL centrifuge tubes.
6. Centrifuge.
7. Filter papers.
8. Separating funnels.

2.5 Microbial Samples

1. *Bacillus subtilis.*
2. *Candida albicans.*

2.6 Screening Endosymbiotic Fungi for Antibacterial Activity

2.6.1 Disc Diffusion Method

1. Sterile paper disks (6 mm in diameter).
2. Micropipette.
3. Muller–Hinton agar (MHA).
4. 30 μg chloramphenicol discs.
5. Ethyl acetate discs.
6. Sterile Petri dishes.
7. Ruler.

2.6.2 Agar Plug Diffusion Method

1. Potato dextrose agar PDA medium.
2. Sterile Petri dishes.
3. Sterile saline.
4. Vortex machine.
5. Muller–Hinton agar (MHA) media.
6. Sterile scapels.
7. Sterile forceps.

8. Parafilm.
9. Incubator.

2.6.3 Agar Dilution Method

1. Agar.
2. Water bath.
3. Streptomycin (3 mg/100 L).
4. Sterile Petri dish plates.
5. 0.85% saline.
6. Marker pen.
7. Micropipette.
8. Incubator.
9. Ruler.

3 Methods

3.1 Collection of Plant Material

1. Collect whole plants at different growth stages.
2. Identify plants using morphology and taxonomic keys.
3. Collect at least five to seven samples of the plants with their roots intact and adhering to soil particles (*see* **Note 1**).
4. Examine the plants and discard the plants where the areas with colorations or lesions are observed. Only healthy material should be used.

3.2 Surface Sterilization

1. Wash the plants under running tap water to remove all the soil particles and surface epiphytes [23], followed by immersion in 70% ethanol for 1 min, then soaking for 4–5 min in sodium hypochlorite (4% active chlorine) and finally three washings for at least 5 min each with demineralized water [24] before drying on a filter paper.
2. Dissect leaves, stems, roots, and flowers using a sterile blade.
3. Gently press the leave, stem, root, and flower samples on filter paper imprinted with Biomalt agar 50 g/L (Biomalt, 120 g/L BactoTM agar, pH 5.6 with antibiotics penicillin G Na 60 mg/L, streptomycin sulfate 80 mg/L, and (Oxy)-tetracycline-HCL 50 mg/L) to check for sterility. This is also a check for surface sterilization. These will serve as control plates (*see* **Note 2**).
5. The final rinse of the plant parts is plated on the same medium, incubated, and observed.

3.3 Isolation and Purification of Endosymbiont Fungi

1. Cut the plant parts into small pieces using sterile scissors and place them in sterile Petri dishes (*see* **Note 3**).

2. Place the plant pieces on aqueous agar (distilled water and 1.5% agar–agar) supplemented with antibiotic streptomycin (3 mg/ 100 L) in Petri dishes and incubate at 28 ± 2 °C for fungal growth. Surface-disinfected and non-disinfected controls should also be placed in the media (*see* **Note 4**).
3. Remove the tips of the fungal hyphae from the aqueous agar (at least 1–2 mm from the edge) and plate on mycological medium, potato dextrose agar (PDA), or Sabourand agar (SA) supplemented with chloramphenicol (0.2 g/L), pH of 7.8, and incubate for 7 days at 27 °C (*see* **Note 5**).
3. Use the colony morphology to examine the fungal culture.
4. Examine and monitor the cultures for 6 weeks for mycelial growth and inoculate on fresh biomalt media.
5. The cultures should undergo purification several times as in **step 4**, before transferring the final pure cultures to PDA slant tubes.
6. The cultures should be grouped according to culture morphology and characteristics.
7. Use the culture morphological characteristics to identify the fungi.
8. Perform phenotypic identification to genus level according to Pitt and Hocking [25] (*see* **Note 6**).
9. Identification to species level may be done using genomic DNA extraction of fungal isolates [26]. Molecular identification is done by DNA amplification and sequencing of the internal transcribed spacer (ITS) region using molecular biological protocols [27].

3.4 Endosymbiont Fungal Cultivation and Extraction

1. Remove four pieces of 5 mm samples using a cork borer from pure colonies of endosymbiont fungi incubated for 3–5 days at a temperature of 25–28 °C.
2. Inoculate four agar plugs in 250-mL Erlenmeyer flasks containing 150 mL of potato dextrose broth (PDB).
3. Cultivate fungi pieces in the fermentation medium for 14 days by shaking using a rotary shaker at 120 rpm.
4. Place the fermented medium into a 15-mL centrifuge tube and centrifuge at 3000 rpm for 20 min.
5. Filter the supernatant using filter paper (*see* **Note** 7).
6. Extract the remaining supernatant with ethyl acetate in a ratio of 1:1.
7. Separate the ethyl acetate extract using a separating funnel from the culture media, then test its primary antibacterial activity.

3.5 Screening Endosymbiotic Fungi for Antibacterial Activity

3.5.1 Disc Diffusion Method

1. Infuse sterile paper disks (6 mm in diameter) with 25 μL of endosymbiont fungi extract, then allow it to dry.
2. Place the discs on the surface of Muller–Hinton agar (MHA) media that has been overgrown with bacteria. The positive control comprises 30 μg chloramphenicol discs, and the negative control discs contain ethyl acetate for the test solution with ethyl acetate [28, 29].
3. Petri dishes containing discs are then incubated at 37 °C for 18–24 h.
4. Measure and calculate the diameter of the inhibition zone formed around the discs.

3.5.2 Agar Plug Diffusion Method

1. Select a well-isolated fungal colony from the growing edge with the aid of a sterile inoculation loop.
2. Spot inoculate the fungal colony at the center of the PDA medium.
3. Incubate the PDA plates for 21 days at 25 ± 2 °C.
4. Select four to five colonies of the target bacteria from a pure culture.
5. Transfer the bacterial colonies into 2 mL of sterile saline and vortex gently.
6. Make an inoculum suspension of 10^8 CFU/mL.
7. Inoculate the bacterial suspension on MHA.
8. Transfer the agar plug by carefully removing growing fungi from the surface of the agar plug using a sterile scalpel.
9. Place the agar plug in the middle of the MHA plate pre-seeded with bacterial suspension using sterile forceps.
10. Seal with parafilm and place at 4 °C for diffusion of metabolites [30].
11. Incubate 37 ± 2 °C for 24 h.
12. Observe and measure the zone of inhibition around the agar plug.

3.5.3 Agar Dilution Method

1. Prepare aqueous agar (distilled water and 1.5% agar–agar) supplemented with antibiotic streptomycin (3 mg/100 L).
2. Allow the sterilized agar to cool to 50 °C in a water bath.
3. Prepare a dilution series of the fungal endosymbionts antimicrobial agents, in 25-mL containers. Include a drug-free control [30].
4. Add 19 mL of molten agar to each container, mix thoroughly, and pour the agar into pre-labeled sterile Petri dishes.

5. Allow the plates to set at room temperature and dry the plates so that no drops of moisture remain on the surface of the agar.
6. Standardize the density of inoculum to give 10^8 colony-forming units (CFU) per spot on the agar.
7. Dilute the suspensions of organisms in 0.85% saline to give 10^7 CFU/mL.
8. Inoculate the plates within 30 min of standardizing the inoculum, to avoid changes in inoculum density.
9. Mark the plates and transfer diluted bacterial suspensions to the wells of an inoculum-replicating apparatus.
10. Use a micropipette to transfer the inocula to the series of agar plates, including a control plate without an antimicrobial agent.
11. Incubate plates at 35–37 °C for 18 h.
12. Measure the zones of inhibition and determine the MIC which is the lowest concentration of the agent that completely inhibits visible growth.

4 Notes

1. The plants may first be photographed for documentation and identification, then placed into a sample bag, which is then placed into a cooler box for further processing [23].
2. The plants are allowed to form impressions on the surface of the medium and then carefully removed using a sterile spatula and discarded).
3. Fragments should be small and thin as colonization may be sparse [31].
4. The selected medium is supplemented with antibiotics to prevent bacterial growth. Antifungal substances may also be added to delay the growth of fast-growing fungal contaminants [32].
5. It is important to include a soaking process [33] if the endophytic fungi are slow-growing. This ensures that intercellular and intracellular endophytic invaders and colonizers of the vascular tissues and root cortex are trapped.
6. Morphological identification should also be done based on aerial mycelia; colony color, growth pattern, and surface texture; margin character; conidial size and color; sporulation; and production of acervuli and hyphae.
7. A portion of the supernatant may be taken and tested for its antibacterial activity.

References

1. Nisa H, Kamili AN, Nawchoo IA, Shafi S, Shameem N, Bandh SA (2015) Fungal endophytes as prolific source of phytochemicals and other bioactive natural products: a review. Microb Pathog 82:50–59
2. Gouda S, Das G, Sen SK et al (2016) Endophytes: a treasure house of bioactive compounds of medicinal importance. Front Microbiol 7:1538
3. Mwamburi LA (2016) Isolation and assessment of stability of six formulations of entomopathogenic *Beauveria bassiana*. In: Glare TR, Moran-Diez ME (eds) Microbial-based biopesticides: methods and protocols. Series: methods in molecular biology, vol 1477. Springer, Humana Press Inc, pp 85–91
4. White JF, Kingsley KL, Zhang Q, Verma R, Obi N, Dvinskikh S et al (2019) Review: endophytic microbes and their potential applications in crop management. Pest Manag Sci 75:2558–2565
5. Xia Y, Sahib MR, Amna A, Opiyo SO, Zhao Z, Gao YG (2019) Culturable endophytic fungal communities associated with plants in organic and conventional farming systems and their effects on plant growth. Sci Rep 9:1669
6. Vega FE (2008) Insect pathology and fungal endophytes. J Invertebr Pathol 98:277–279
7. Estrada C, Wcislo WT, Van Bael S (2013) Symbiotic fungi alter plant chemistry that discourages leaf-cutting ants. New Phytol 198: 241–251
8. Vidal S, Jaber LR (2015) Entomopathogenic fungi as endophytes: plant-endophyte-herbivore interactions and prospects for use in biological control. Curr Sci 109:46–54
9. Mwamburi LA (2021) Endophytic fungi, *Beauveria bassiana* and *Metarhizium anisopliae*, confer control of the fall armyworm, *Spodoptera frugiperda* (J. E. Smith) (Lepidoptera: Noctuidae), in two tomato varieties. Egypt J Biol Pest Co 31:7
10. Chaudhary P, Agri U, Chaudhary A, Kumar A, Kumar G (2022) Endophytes and their potential in biotic stress management and crop production. Front Microbiol 13:933017
11. Ghaffari MR, Mirzaei M, Ghabooli M, Khatabi B, Wu Y, Zabet-Moghaddam M et al (2019) Root endophytic fungus *Piriformospora indica* improves drought stress adaptation in barley by metabolic and proteomic reprogramming. Environ Exp Bot 157:197–210
12. Qiang X, Ding J, Lin W, Li Q, Xu C, Zheng Q et al (2019) Alleviation of the detrimental effect of water deficit on wheat (*Triticum aestivum* L.) growth by an indole acetic acid-producing endophytic fungus. Plant Soil 439: 73–391
13. Bajaj R, Huang Y, Gebrechristos S, Mikolajczyk B, Brown H, Prasad R et al (2018) Transcriptional responses of soybean roots to colonization with the root endophytic fungus *Piriformospora indica* reveals altered phenylpropanoid and secondary metabolism. Sci Rep 8:10227
14. Aslam MM, Karanja J, Bello SK (2019) *Piriformospora indica* colonization reprograms plants to improved P-uptake, enhanced crop performance, and biotic/abiotic stress tolerance. Physiol Mol Plant Pathol 106:232–237
15. Chepsergon J, Mwamburi L, Kiprop E (2016) Ability of *Trichoderma harzianum* from semi-arid soils to enhance antioxidant defense of maize seedlings under water stress. Br Microbiol Res J 17:1–9
16. Saikkonen K, Wäli P, Helander M, Faeth SH (2004) Evolution of endophyte–plant symbioses. Trends Plant Sci 9:275–280
17. Bae H, Roberts DP, Lim H-S, Strem MD, Park S-C, Ryu C-M et al (2011) Endophytic *Trichoderma* isolates from tropical environments delay disease onset and induce resistance against *Phytophthora capsici* in hot pepper using multiple mechanisms. Mol Plant-Microbe Interact 24:336–351
18. Mwamburi LA (2020) Beauveria. In: Amaresan N, Senthil Kumar M, Annapurna K, Kumar K, Sankaranarayanan A (eds) Beneficial microbes in agro-ecology: bacteria and fungi. Elsevier, Academic Press, pp 727–748
19. Mwikali P, Mwamburi L, Musinguzi SP (2024) Formulation and bio-efficacy of different isolates of *Beauveria bassiana* against adults and third nymphal instar of desert locust (*Schistocerca gregaria* Forskål). Biol Control 198: 105642
20. Mwikali P, Mwamburi L, Musinguzi SP (2025) Biological characteristics and pathogenicity of entomopathogenic *Beauveria bassiana* (Bals.-Criv.) Vuill (Hypocreales: Cordycipitaceae) isolated from soils in northern Kenya against desert locust (*Schistocerca gregaria*) (Forskål) (Orthoptera: Acrididae). Crop Prot 190: 107126
21. Jia M, Chen L, Xin H-L, Zheng C-J, Rahman K, Han T, Qin L-P (2016) A friendly relationship between endophytic fungi and medicinal plants: a systematic review. Front Microbiol 7:906

22. Mwamburi LA (2023) Cultivation-based studies of endophytic fungi in plants. In: Sankaranarayanan A, Amaresan N, Dwivedi MK (eds) Endophytic microbes: isolation, identification, and bioactive potentials. Springer protocols handbooks. Humana, New York, pp 145–151
23. Mwamburi LA (2025) Collection and processing of plant fungal endophytes from shrubs. In: Dharumadurai D, Narayanan AS (eds) Plant microbiome engineering. Methods and protocols in food science. Humana, New York, pp 19–25
24. Mat-Jalil M, Ibrahim D (2018) Fungal endophytes isolated from the leaves of a medicinal plant, *Ocimum sanctum* Linn and evaluation of their antimicrobial activities. Afri J Microbiol Res 12:616–622
25. Pitt JI, Hocking AD (2009) Introduction. In: Fungi and food spoilage. Springer, Boston, pp 1–2
26. Nicole P, Kwiatkowskwi NP, Babiker WM, Merz WG, Carroll KC, Zhang SX (2012) Evaluation of nucleic acid sequencing of the D1/D2 region of the large subunit of the 28S rDNA and the internal transcribed spacer region using smartGene IDNS software for identification of filamentous fungi in a clinical laboratory. J Mol Diagn 14:393–401
27. Schoch CL, Seifert KA, Huhndorf S, Robert V, Spouge JL, Levesque CA, Chen W (2012) Fungal barcoding consortium. Nuclear ribosomal internal transcribed spacer (ITS) region as a universal DNA barcode marker for fungi. Proc Natl Acad Sci USA 109:6241–6246
28. Chemweno T, Mwamburi L, Korir R, Mutuku A, Bii C (2018) Antimicrobial activity and safety of *Maesa lanceolata* for the treatment and management of selected bacterial pathogens. J Adv Microbiol 8:1–8
29. Mutuku A, Mwamburi L, Keter L et al (2020) Evaluation of the antimicrobial activity and safety of *Rhus vulgaris* (Anacardiaceae) extracts. BMC Complement Med Ther 20:272
30. Balouiri M, Sadiki M, Ibnsouda SK (2016) Methods for *in vitro* evaluating antimicrobial activity: a review. J Pharm Anal 6:71–79
31. Anjum N, Chandra R (2019) Endophytic bacteria of *Catharanthus roseus* as an alternative source of vindoline and application of response surface methodology to enhance its production. Arch Biol Sci 71:27–38
32. Kjer J, Debbab A, Aly AH, Proksch P (2010) Methods for isolation of marine-derived endophytic fungi and their bioactive secondary products. Nat Prot 5:479–490
33. Sieber T (2002) Fungal root endophytes. In: The Hidden Half, 4th edn. CRC Press, Taylor & Francis Group, Boca Raton, pp 887–917

Chapter 22

Screening of Endosymbionts for Anticancer Compounds

Satish V. Patil, Atharva S. Patil, Vikas S. Patil, and Jitendra D. Salunkhe

Abstract

The current statistical data on mortality and morbidity cases of cancer indicates that there is a worldwide threat of different types of cancers. On this background, there is an urgent need to search for an effective and safe anticancer agent in order to avoid drawbacks of anticancer agents like side effects, ineffectiveness, and cost. Searching for natural sources of anticancer metabolites is one of the most promising ways in current situations. Among these natural sources like plants, animals, and microbes, the exploitation of microbial metabolites has proven to be a significant source for novel metabolites due to the vast diversity of microbes. Microbial endosymbionts are organisms which play a key role in plant physiology and are also characterized for producing potent metabolites.

To enhance the rate of searching such anticancer producers, there is simply a need for primary screening. Applications of the Ames test are proposed as primary screening of microbial metabolites and further forwarded for confirmation by simple MTT assay and cytotoxicity assay.

Key words Anticancer, Ames' test, Cytotoxicity, MTT assay, Metabolites

1 Introduction

Cancer disease is one of the major causes of mortality globally. Worldwide annually, different cancer cases are ever increasing steadily. According to WHO and the International Agency for Research on Cancer (IARC), the recent burden of cancer is ever-increasing; in 2022, 9.7 million deaths and 20 million new cancer cases globally were estimated. It was also estimated that among various cancer forms, the major contributions were the cases related to the lung, breast, and colorectal. The global effort cost for developing anticancer is about US$ 100 billion; however, the currently available anticancer drugs are not so effective and have adverse side effects like toxicity on nontarget cells [1, 2]. These all-estimated figures indicate a need to search for new effective anticancer agents from natural sources such as microbes, plants, and animals [2, 3].

N. Amaresan and Sapna Chandwani (eds.), *Endosymbionts*, Springer Protocols Handbooks,
https://doi.org/10.1007/978-1-0716-4771-4_22,

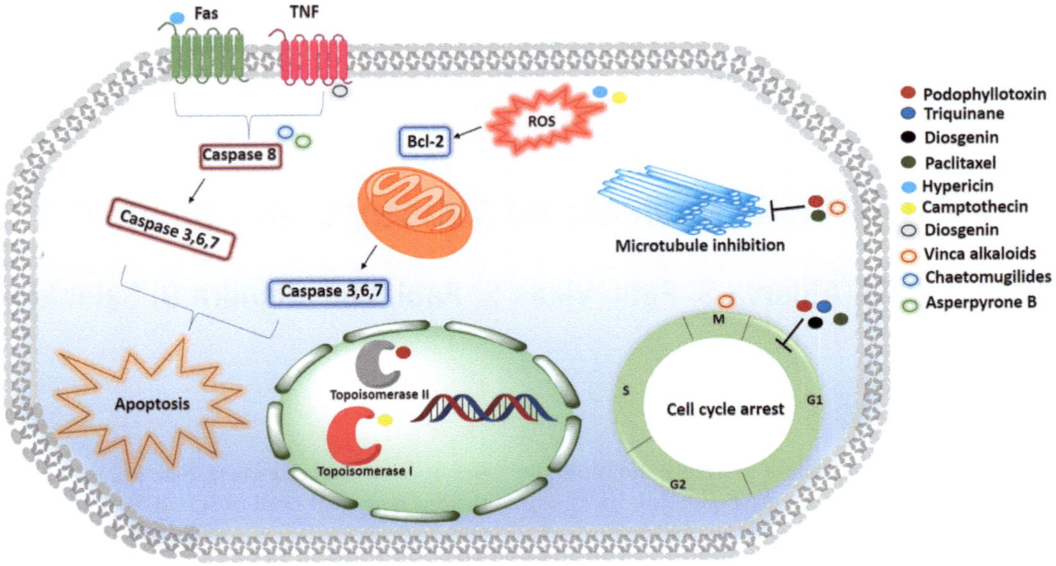

Fig. 1 Known mechanisms of natural anticancer agents for induction of apoptosis

Among these natural sources, the microbial source has more potential and is considered a future candidate for novel effective anticancer sources because of its vast diversity with a diverse mechanism of action which induces apoptosis of cancer cells (Fig. 1). Among the microbes, endosymbiont fungi are reported as potent source of antioxidant, anticancer, antimicrobial, and bioactive metabolites [4]. The effective anticancer agent is produced from a plant part, i.e., *Taxus brevifolia* commonly known as the Pacific yew tree, but the tremendous requirement of active compound is not fulfilled by this source, as the plant source metabolite production also depends on various factors like drought, diseases, variety, and soil. On this background, Hao et al. [5] advocated the exploitation of endophytic fungi. There are various fungi that have been reported for their potential to produce taxol, i.e., paclitexe which includes *Fusarium solani*, *Cladosporium oxysporum*, *Beauveria* sp., *Alternaria alternata*, *Colletotrichum gloeosporioides*, *Glomerella cingulata*, and *Nigrospora sphaerica* [6–8]. It was proved that these metabolites induce apoptosis of cancer cells. Besides these, some specific fungi, *Myrothecium roridum*, produce *Myrotheciumone* A, *Cladosporium* produces metabolite Cladodosporol H, *Chaetomium globosum* produces metabolite *Antioquenian* B9, and *Aspergillus fumigatus* produces podophyllotoxin which were characterized for its anticancer potential [9–12]. Primary screening of anticancer metabolites is important which is primarily carried out by two major assays: firstly by the application of the Ames test and, secondly, by testing against the cell lines via in vitro cell viability assay.

2 Materials

2.1 Primary Screening Ames' Test

1. Standard *E. coli* or *S. typhimurium* strains, e.g., TA98, TA1535, and A1537.
2. Minimal glucose medium butts.
3. Sterile Petri plates.
4. 0.1% DMSO.
5. Phosphate buffer (0.2 M, PH 7.0).
6. Extracted test sample—endosymbiotic fungi or bacterial metabolites.
7. 3-(4,5-Dimethylthiazol-2yl)-2,5-diphenyl tetrazolium bromide (MTT).
8. As standard mutagens: 2-nitrofluorene, 4-nitroquinoline N-oxide, sodium azide, 9-aminoacridine, T 2-aminoanthracene, etc.

2.2 Cytotoxicity Assay/MTT Assay

Dissolve 5 mg 3-(4,5-dimethylthiazol-2yl)-2,5-diphenyl tetrazolium bromide (MTT) in 1 mL 1× PBS. Sterilize by filtration.

2.2.1 MTT Reagent

1. Dimethylsulfoxide (DMSO)/0.2% nonidet P-40 (NP-40) (*see* **Note 1**).
2. 96-well sterile polypropylene plate.
3. Cell cultures, normal cell line, MCF10A, HepG2 (human hepatocellular carcinoma cell line), HCT116 (colorectal adenocarcinoma), MCF-7 (breast adenocarcinoma cell line).
4. Roswell Park Memorial Institute (RPMI) medium or Dulbecco's Modified Eagle Medium (DMEM) with 10% fetal bovine serum and 1% penicillin/streptomycin (*see* **Notes 2–4**).

3 Methods

3.1 Extraction of Metabolites

1. Separate biomass by centrifugation at 10,000 rpm, from the fermentation medium comprising of endosymbionts fungi or bacteria.
2. Extract using polar and nonpolar solvents, and separate as well as purify as per the nature of the compound.
3. Lyophilize the extract and further use it as a test compound for anticancer screening including primary and confirmation assay.

3.2 Primary Screening

1. Grow the standard strains in the 50 mL of nutrient broth/ Luria broth with a small amount of histidine.
2. After incubation of 24 h, centrifuge at 1000 rpm, collect cell mass, and suspend in sterile 50 mL phosphate buffer, pH 7.0.

3. Prepare three tubes labeled as a positive control, test sample as control, and test sample with standard mutagen.
4. For positive control, add 100 μL of aminofluorene or available standard mutagen in 5 mL buffered suspensions and label it as positive control (*see* **Note 1**).
5. In 5 mL buffered cell suspension, add 100 μL of test anticancer metabolites extracted from microbial broths and labeled as control.
6. Mix 100 μL of aminofluorene and 1000 μL of microbial test metabolites in 5 mL buffered cell suspension and label it as a test compound.
7. Incubate above three tubes at 37 °C for 20 min.
8. After incubations, prepare sterile six minimal agar medium butts.
9. Add 100 μL of treated buffered bacterial suspensions in minimal agar butts at 40 °*C*, mix well, and pour it into a sterile Petri plate aseptically. Mix well by rotating plates clockwise and anticlockwise, and then, incubate for 48 h at 37 °C.
10. After incubation, count the colonies on each plate.
11. Compare the number of colonies in positive control, test control, and mix metabolites and standard mutagen treatment.
12. If the average number of grown cells on mix metabolites and standard mutagen treatment is less than the standard mutagen, it primarily indicates that the tested microbial metabolites have anticancer potential.

3.3 Confirmation of Anticancer Metabolites by MTT Assay

1. Further confirmations are made by the cytotoxicity assay.
2. Grow a specific cancer cell line separately on the RPMI medium or DMEM with 10% fetal bovine serum and 1% penicillin/streptomycin.
3. Grow normal cell line, MCF10A, separately on RPMI medium or DMEM with 10% fetal bovine serum and 1% penicillin/streptomycin.
4. Dissolve pure microbial metabolites in 1 mL of 0.1% DMSO-prepared concentrations 10–200 μg/mL and add 50 μL in 96-well plates in triplicate.
5. Add 50 μL (10 μg/mL) Taxol and use it as a positive control and only RPMI with the cell as a negative control.
6. Wash the confluent growth of the grown cell line with washing buffer.
7. Dilute with the RPMI and add 50 μL of cell suspension in 90 wells in triplicate including negative control and Taxol containing positive control.

8. Incubate for 3 h and observe common signs such as rounding shrinkage of cell layer and loss of layer.
9. Add 10 μL of MTT solution in each well, shake it at 120 rpm on a shaker, and incubate it in a CO_2 incubator for 2 h.
10. Remove the liquid layer above the formazan crystal and then add 100 μL DMSO (*see* **Notes 5–8**).
11. Measure the absorbance at 570 nm (*see* **Note 9**).
12. Calculate by formula IC_{50} = (absorbance of the cell without treatment − absorbance of cells with treatment/absorbance of the cell without treatment) × 100.
13. Selective index (SI) = CC50 value for normal cells/IC_{50} for cancer cells.

4 Notes

1. Use standard limit concentrations for chemical mutagen in control as below:
   Aminoacridine 3 μg/mL, 2-aminoanthracene 0.4 μg/mL, 4-nitroquinoline N-oxide 0.04 μg/mL, aminoanthracene 0.4 μg/mL.
2. Dulbecco's Modified Eagle Medium is a widely used basal medium to grow various cell lines. A readily available dehydrated medium is available by various companies.
3. RPMI-1640 medium is the popularly used medium for growing various cell lines tissue, developed by Roswell Park Memorial Institute [13].
4. Keep 100 μL DMSO in a medium without cells as a blank.
5. From each well, remove media carefully without disturbing cells.
6. Mix DMSO to each well by using a sterile micropipette up and down.
7. For some cell lines, check DMSO cytotoxicity.
8. 0.2% nonidet P-40 (NP-40) or a substitute with 8 mM HCL in isopropanol is an alternative to DMSO for dissolving formazan crystals.
9. Measure the absorbance at 570 nm immediately.

Acknowledgement

Prof. Dr. Satish V Patil kindly acknowledges Prof. Beom Soo Kim, South Korea, Chemical Engineering, Chungbuk National University, for his constant support and inspiration.

References

1. Prasad V, De Jesús K, Mailankody S (2017) The high price of anticancer drugs: origins, implications, barriers, solutions. Nat Rev Clin Oncol 14:381–390
2. Uzma F, Mohan CD, Hashem A, Konappa NM, Rangappa S, Kamath PV et al (2018) Endophytic fungi alternative sources of cytotoxic compounds: a review. Front Pharmacol 9: 309
3. Remesh A (2017) Toxicities of anticancer drugs and its management. Int J Basic Clin Pharmacol 1:2–12
4. Strobel GA, Hess WM, Li JY, Ford E, Sears J, Sidhu RS et al (1997) *Pestalotiopsis guepinii*, a taxol-producing endophyte of the wollemi pine, *Wollemianobilis*. Aust J Bot 45:1073–1082
5. Hao X, Pan J, Zhu X (2013) Taxol producing fungi. In: Ramawat K, Mérillon JM (Eds.) Natural products Springer, Berlin 2797–2812
6. Chakravarthi BV, Das P, Surendranath K, Karande AA, Jayabaskaran C (2008) Production of paclitaxel *by Fusarium solani* isolated from Taxus celebica. J Biosci 33:259–267
7. Gokul R, Manikandan K, Arulvasu R, Pandi M (2015) Anti-proliferative effect of fungal taxol extracted from *Cladosporium oxysporum* against human pathogenic bacteria and human colon cancer cell line HCT 15. Spectrochim Acta A Mol Biomol Spectrosc 138:667–674
8. de Andrade HF, de Araújo LCA, dos Santos BS, Paiva PMG, Napoleão TH, dos Santos Correia MT et al (2018) Screening of endophytic fungi stored in a culture collection for taxol production. Braz J Microbiol 49:59–63
9. Lin T, Wang GH, Shan WJ, Zeng DQ, Ding R, Jiang X (2014) Myrotheciumones: bicyclic cytotoxic lactones isolated from an endophytic fungus of *Ajuga decumbens*. Bioorg Med Chem Lett 24:2504–2507
10. Tantapakul C, Promgool T, Kanokmedhakul K, Soytong K, Song J, Hadsadee S et al (2020) Bioactive xanthoquinodins and epipolythiodioxopiperazines from *Chaetomium globosum* 7s-1, an endophytic fungus isolated from *Rhapis cochinchinensis* (Lour.) Mart. Nat Prod Res 34:494–502
11. Li HL, Li XM, Mandi A, Antus S, Li X, Zhang P et al (2017) Characterization of cladosporols from the marine algal-derived endophytic fungus Cladosporium cladosporioides EN-399 and configurational revision of the previously reported cladosporol derivatives. J Org Chem 82:9946–9954
12. Xianzhi Y, Shiping G, Lingqi Z, Hua S (2003) Select of producing podophyllotoxin endophytic fungi from podophyllin plant. Nat Prod Res Develop 15:419–422
13. Moore GE, Gerner RE, Franklin HA (1967) Culture of normal human leukocytes. JAMA 199(8):519–524

Chapter 23

Screening of Endosymbionts for Nematicidal Activity

Neelam Thakur, Samiksha Jhamta, and Simranjeet Kaur

Abstract

A long-term biocontrol method against parasitic nematodes is provided by endosymbionts, such as symbiotic bacteria like *Xenorhabdus* and *Photorhabdus* and entomopathogenic nematodes (EPNs). In this work, endosymbiotic bacteria and fungi are isolated, and their nematicidal activity is screened out. Bacteria are isolated from infected insects, and samples of soils and roots are gathered and exposed to baiting procedures for the purpose of isolating EPNs. On selective media, fungi are cultivated from soils and root samples. Nematicidal efficacy is examined using in vitro and in vivo tests, as well as testing on susceptible plant kinds conducted in polyhouse. The outcomes showed a notable decrease in the need for chemical nematicides and a notable increase in nematode mortality, highlighting the potential of endosymbionts as environmentally friendly biocontrol agents in sustainable agriculture.

Key words Biocontrol, Endosymbionts, Nematicidal activity, Plant-parasitic nematodes, Sustainable agriculture

1 Introduction

Plant-parasitic nematodes (PPNs) are microscopic roundworms that significantly damage various crops such as soybeans, capsicum, corn, potatoes, tomatoes, rice, and carrots [1–5]. Among them, the genera *Heterodera*, *Globodera*, *Meloidogyne*, and *Pratylenchus* can cause significant agricultural losses, affecting food security and economic stability [2, 6]. Endosymbionts are microorganisms that reside within the tissues of a host organism, often offering mutual benefits [7]. Endosymbionts play a crucial role in managing nematodes. They can create bioactive substances that are poisonous to nematodes, serving as organic biocontrol agents [8]. Entomopathogenic nematodes (EPNs) and endosymbiotic bacteria like *Xenorhabdus* and *Photorhabdus* have demonstrated a great deal of promise for managing nematode populations [8, 9].

These endosymbionts not only help in reducing the reliance on chemical nematicides, which can have harmful environmental effects, but also promote sustainable agricultural practices

N. Amaresan and Sapna Chandwani (eds.), *Endosymbionts*, Springer Protocols Handbooks,
https://doi.org/10.1007/978-1-0716-4771-4_23,

[8, 10]. Through the utilization of endosymbionts' inherent antagonistic characteristics, scientists hope to create environmentally sustainable methods of controlling PPNs [11, 12]. Endosymbionts have attracted a lot of attention in the agricultural setting because of their potential to act as biocontrol agents against parasitic nematodes [13, 14].

Culturing endosymbiotic bacteria and fungi allows researchers to study their biology, genetics, and interactions in controlled environments. Incorporating endosymbionts into nematode management measures is crucial, as concerns regarding the usage of chemical pesticides and the need for sustainable farming practices continue to rise [15–17]. This strategy not only guarantees efficient nematode management but also fosters a more robust and long-lasting agricultural ecology.

2 Materials

A vital first step in creating biocontrol techniques for managing plant-parasitic nematodes is isolating endosymbiont bacteria from the soil surrounding plant roots. This is a broad synopsis of the procedure.

2.1 Sample Collection

1. Soil Sampling: Collect soil samples from the area of the soil near the rhizosphere, as a lot of microbiological activity takes place there (*see* **Note 1**).
2. Root Sampling: Gently dig up plant roots and gather them with attached soil (*see* **Note 2**).

2.2 Isolation of Microbes

2.2.1 Materials Needed

Sterilized polythene bags, sterilized scalpel and sterilized soil auger, plastic containers 250 g, sterilized motor and pestle, sterilized distilled water (SDW), sodium hypochlorite solution (NaOCl) (1%), agar powder (AP), nutrient agar (NA), nutrient bromothymol blue tetrazolium chloride agar (NBT agar) (peptone (5 g) + sodium chloride(5 g) + beef extract powder (2.5 g) nutrient broth 13.0 g + agar 20.0 g + bromothymol blue (BTB) 0.025 g + triphenyl tetrazolium chloride (TTC) 0.04 g + water), ampicillin (100 μg/mL), cholesterol (10 μg/mL), potato dextrose broth (PDB), sterilized spreader, Petri plates, sterile needle, micropipettes, biological oxygen demand (BOD) incubator, shaker, laminar air flow, autoclave, centrifuge, ethanol 70%, glycerol 25%, glass slides, conical flask, Neubauer haemocytometer (NH), TE buffer (10 mM Tris HCL (1.576 g) + EDTA (0.372 g) + 1000 mL distilled water), SET buffer (75 mMNaCl (5.844 g) + 25 mM EDTA (9.306 g) + 20 mM Tris base (3.152 g) + 1000 mL distilled water), PCR (genomic DNA (1 μL) + Taq buffer (10×) (2.0 μL) + $MgCl_2$ 25 mM (1.2 μL) + dNTPs (1 mM) (1.0 μL) + primer forward

(1.0 μL) + reverse primer (1.0 μL) + Taq polymerase (1.0 μL)), agrose gel (0.8% and 1.2%), electrophoresis, gel documentation (Alpha-Imager), Stereozoom microscope (Leica S9i) and compound microscope (Leica DM750) (*see* **Notes 5** and **6**).

2.3 Nematicidal Efficacy of Endosymbionts

2.3.1 Materials Needed

Bacterial or fungal culture, shaker, incubator, centrifuge, millipore filter, micro-centrifuge tube, microplate plates, SDW, population of *Meloidogyne incognita* and other parasitic nematodes (infected juvenile IJ2 suspension), susceptible plant variety, polyhouse (26 ± 2 °C with 15–16 h light availability), eggs of the *M. incognita* of any other nematodes, NaOCl solution (15–17 g per liter), extraction trays, agar discs, conical flask 250 mL, filter paper, Petri plates, fine mesh screen (0.15 mm), soil mixture (loam:sand = 1:1), pots (size 35–45 cm), surface-sterilized plant seeds, fertilizers (N + P + K = 14: 10:14); computer with SPSS 25.0 software (*see* **Notes 3** and **4**).

3 Methods

3.1 Soil Baiting Technique for Isolation of Entomopathogenic Nematodes

1. As bait, use the larvae of the rice moth (*Corcyra cephalonica*) or the wax moth (*Galleria mellonella*) [20].
2. For a period of 5 to 7 days, place the soil samples in plastic containers with the bait insects and maintain them in a controlled atmosphere.
3. Infected and killed bait insects confirm the nematode presence. Then, use a modified white trap to hold infected bait insects in moist Petri plates [19].
4. Gather the dead insects and further, for additional research, extract the nematodes [20, 21]. Further, use the molecular methods and physical traits to identify the isolated nematode (*see* **Note 10**).

3.1.1 Bacterial Isolation

1. Symbiotic bacteria are present in dead insects that have been infected by EPNs. To extract the nematodes, the infected insects are now dissected in sterile circumstances [18].
2. In their stomach, the nematodes have bacteria called *Photorhabdus* and *Xenorhabdus*. Using molecular methods and physical traits, identify the isolated bacteria (Fig. 1).
3. Using a sharp needle, remove 100 infectious juveniles of EPNs. To create pellets, centrifuge the collected IJs for 5 min at 5000 rpm. Surface sterilize this pellet for approximately 5 minutes by dissolving it in 1% sodium hypochlorite solution.
4. Next, add 1 cc of distilled water for washing, and centrifuge the mixture once more for 5 min at 5000 rpm. Crush the IJs using an autoclaved sterile mortar and pestle and then use a sterile needle to streak a loop of the crushed culture on NBT agar

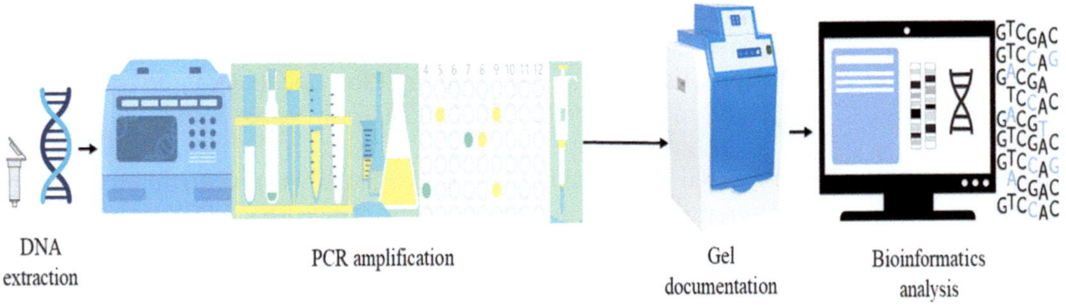

Fig. 1 Experimentation setup for isolation of symbiotic fungi

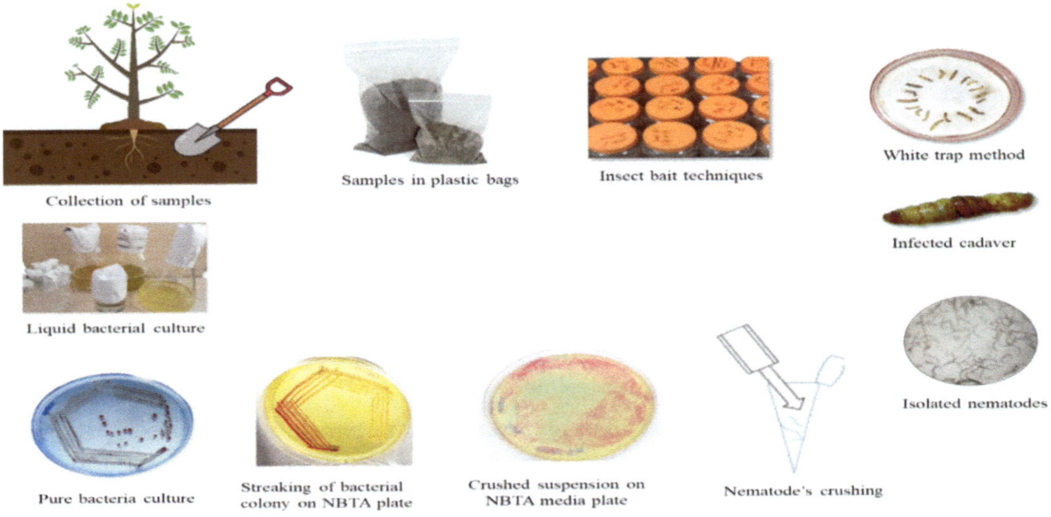

Fig. 2 Experimentation setup for isolation of symbiotic bacteria

media in Petri plates. Incubate the Petri plates at 28 ± 1 °C for 48 h [19] (*see* **Notes 11** and **12**).

3.1.2 Culture Media

1. Perform re-streaking on the same medium as mentioned above and incubation to purify the single bacterial colonies [22].
2. Use the NBT agar medium to streak and purify the bacterial culture that is isolated from individual colonies, and then culture them at 28 ± 1 °C. Keep the plates exhibiting bacterial growth at 41 °C and designate them as master plates after 48 h.
3. For later usage, the fully developed bacterial colonies are stored in stock containing 25% glycerol and NB (Fig. 2) (*see* **Note 13**).

3.1.3 Fungal Isolation

Collect soil and root samples from plants, sterilize them, place them on selective media, incubate them for 5 to 7 days at a temperature between 25 °C and 28 °C, extract and purify the fungi, and identify the isolated fungi using molecular and morphological methods.

Samples are taken from various areas, the roots are sterilized, and they are then sub-cultivated on new media.

3.2 Culture Media for Isolated Fungus

1. Potato dextrose agar (PDA) is prepared by combining agar and dextrose with potato extract. To make 100 mL of material, 20 g each of potato extract, dextrose, and agar are combined and poured into a conical flask along with distilled water.
2. After autoclaving, the conical flask is sealed with a cotton plug. The autoclaved material is then added to a Petri plate and allowed to settle.
3. Each of the three autoclaved test tubes contains 9 mL of distilled water. In the first test tube, 2×10^4 spores/mL of the isolated fungus is added and mixed thoroughly.
4. Using a pipette, extract 1 mL of the solution from the first test tube, and mix it properly in the second test tube.
5. Now, using a micropipette, take 100 μL, or 1×10^4 spores/mL, of the solution from the third test tube and place it onto a Petri plate containing PDA.
6. Spread the solution using a spreader and place the Petri plate inside an incubator to grow. Identify fungi from the mother plates, purify them on PDA plates, and then place them in an incubator to continue growing.

3.3 Nematicidal Efficacy of Endosymbiotic Microbes

Soil, roots, and nematodes are the sources of endosymbiotic microorganisms such as bacteria and fungi. After cultivation, they undergo laboratory testing against plant-parasitic nematodes and a greenhouse test. Field testing is the next step after successful trials, evaluating the plant's potential as a biocontrol agent and supporting sustainable farming methods.

3.3.1 Endosymbiotic Fungus

1. Agar cakes from 14-day-old PDA plates are inoculated into 200 mL PDB using Erlenmeyer flasks (250 mL), and the mixture is incubated at 26–28 °C for 6–7 days at 100 rpm.
2. 1 mL of liquid medium is added to the micro-centrifuge along with 100 hatched infected second-stage juveniles (IJs) of *M. incognita*, and the mixture is centrifuged for 1 min at 2500 rpm. At 28 °C, the obtained supernatant is cultured.
3. In the greenhouse, a host plant variety that is extremely vulnerable to *M. incognita* infections is utilized. Using sterile forceps, ten egg masses are randomly removed from the plant root system and placed in sterile water.
4. After rinsing the egg masses and IJs three to four times with sterile water, place them on PDA. There are clearly identifiable colonies based on their size, shape, and color. Fungal culture filtrates (FCFs) are obtained using Whatman No. 1 filter paper

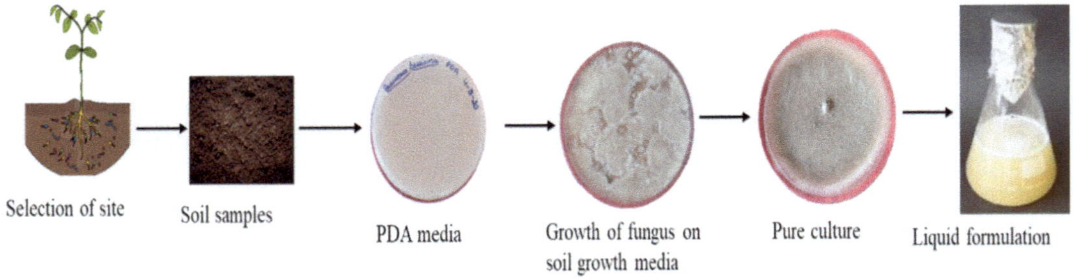

Fig. 3 Experimentation setup for isolation of symbiotic fungi

and a millipore filter (0.22 μm). FCF is prepared in a 250-mL flask containing 100 mL of PDB medium (pH 5.0) for the fungus culture.

5. The flask is subsequently shaken at 150 rpm for 7 days at 28 ± 2 °C in the dark. Whatman No. 1 filter paper is then used twice to filter the culture broth.
6. To exclude fungal spores, the culture broths are passed through a 0.45-μm membrane filter. The filtrates are applied at the initial preparation concentration. As previously noted, FCF is obtained by separating the liquid from the mycelia (Fig. 3).
7. 5 mL of FCF and 1 mL of IJ2 solution (100 IJs/mL) are pipetted into each Petri dish to test the FCF's potential toxicity (in vitro) against IJs and egg masses.
8. Pipette 1 mL of IJ suspension and 5 mL of PDB into the control Petri plates. After 48 h, the percentage of nonmotile nematodes is determined.
9. For 5 h, they can be revived in tap water. Dead nematodes are identified as those that show no signs of life. For both the treatment and control groups, a total of five replicates are employed.
10. Using SPSS's ANOVA and the least significant difference (LSD) post hoc test, significant differences are determined (*see* **Note 8**).

3.3.2 Screening of Endosymbiont Fungi for Nematicidal Activity (In Vivo)

1. For 15 days, the isolated fungi are cultured on PDA plates at 28 ± 2 °C. After that, 5 mL of sterile water is poured over the plate surface, and the spores and conidia are removed from the Petri plates.
2. The fungal slurry is filtered through a fine mesh screen (0.15 mm) to separate the spores from the hyphae. Using a Neubauer hemocytometer, the concentration of the spore suspension is adjusted to 1×10^7 spores per mL, and it is stored at 4 °C for later use.

3. After autoclaving the soil substrate combination for 2 h at 121 ° C, it is transferred into 35–45 cm pots. The soil is well mixed with 3 mL of IJ2 suspension.
4. Surface sterilized plant seeds are buried 1 cm deep in the soil mixture. One milliliter of the tested isolate suspension is pipetted on top of the seed, and topsoil is then applied.
5. Next, the control seeds are pipetted with 1 mL of water. For every fungal endosymbiont, five replicates in total—including the control sets—were created.
6. The plants are kept in the polyhouse, and daily water is provided for better growth. Additionally, they are manured every 6 days using the mentioned ratio of 3–4 g per liter of fertilizer.
7. The pots are arranged using a technique called completely randomized design (CRD). After 5 weeks of seed germination, the plants are harvested, and the number of infected lesions or other signs, such as galls, is counted (*see* **Note 8**).
8. Using SPSS's ANOVA and the least significant difference (LSD) post hoc test, significant differences are determined [23] (*see* **Notes 14** and **15**).

3.3.3 Endosymbiotic Bacteria

1. Chen and Dickso [24] examined the following in vitro test: Each well in 24-well microtiter plates receives 200 μL of bacterial suspension (10^9 CFU/mL) in SDW. Fifty milliliters of surface-sterilized nematode suspension in SDW is then added to each well, containing 30 nematodes.
2. Every treatment is carried out five times. Wells that contain solely SDW are regarded as the controls.
3. Incubation for the plates took place at 28 °C. After adding 20 mL of 1 N NaOH, the number of live and dead nematodes is counted under a stereo microscope after 48–72 h.
4. The percentage mortality of the nematodes relative to the control is computed, and the results are documented.
5. After angular transformation, the data are analyzed using ANOVA (*see* **Note 8**).
6. In 100 mL of NB culture medium, the refined bacterial strains are incubated for 24 h at 160 revolutions per minute at 28 °C.
7. The bacterial growth is then centrifuged for 5 minutes at 8000 rpm. The cells are now taken out of the membrane filter in the third step.
8. A micro-centrifuge tube holding 1 mL of liquid medium is filled with 100 hatched pre-parasitic second-stage juvenile nematodes, and the tube is centrifuged for 1 minute at 2500 rpm.

9. The obtained supernatant is kept in an incubator set at 28 °C. A very susceptible kind of host plant is selected to be kept in the polyhouse with a population of *M. incognita* or any other nematodes. Fifteen grams of NaOCl per liter solution is used to extract nematode eggs from damaged plant roots.
10. To produce infected juveniles, nematode eggs were cultured in extraction trays at 25 °C for 14 days. The IJ suspension is accustomed at 100 nematodes/mL (*see* **Note 16**).

4 Notes

1. Soil and root samples are crucial for evaluating nematicidal properties due to their diverse microbial diversity, nematode proximity, and rich organic compounds, which attract and nourish microorganisms, enhancing bioactive compound production.
2. To prevent contamination that could skew the results, make sure all sampling instruments and containers are sterile. To guarantee a representative microbial population for screening, gather enough soil and root samples. Sample when plants are actively growing, as this is when the rhizosphere's microbial activity is most likely to be at its peak.
3. To ensure accurate record keeping and traceability, properly label all samples with pertinent information (such as location, date, and plant species). To maintain microbial viability during storage and transportation, keep samples cold. To guarantee ideal circumstances for nematode infection and emergence, maintain a regulated environment throughout incubation.
4. To prevent cross-contamination, dissect contaminated insects in a sterile environment. To accurately identify bacteria and nematodes, use both morphological and molecular approaches. To isolate pure bacterial cultures, make sure nematodes are effectively surface sterilized.
5. To encourage the growth of the target bacteria, closely monitor the incubation conditions. Sterilize samples correctly to get rid of impurities and encourage the growth of the desired fungus. To facilitate fungal development, make sure PDA and other media are accurately prepared. To get the required fungal spore concentration, precisely perform successive dilutions.
6. To maintain microbial viability throughout sample transportation and storage, maintain the proper temperatures. To guarantee traceability and prevent confusion, accurately mark all samples and plates.

7. Adhere to laboratory safety procedures, such as the use of personal protective equipment (PPE) and disposing of biological waste appropriately. Researchers performing comparable studies will find complete help in this detailed procedure and note section.
8. To guarantee statistical correctness, perform enough duplicates. Constantly add benchmark treatments to compare against control.
9. Take great care when handling any biological material to prevent damage or viability loss.
10. To use the soil baiting approach, soil samples are exposed to nematode-infecting insect larvae, most often wax moth larvae (*Galleria mellonella*). Nematodes are then separated from infected larvae after they have been gathered. This approach is a dependable way to separate EPNs from various soil habitats because it replicates the natural circumstances that nematodes would come across and infect their insect hosts.
11. To fully utilize the advantageous qualities of microorganisms for environmentally conscious farming methods, the development of new bioactive chemicals, and sustainable agricultural practices, it is essential to separate bacteria and fungus from soil to evaluate their endosymbiotic potential.
12. In microbiology, morphological identification is still an essential tool. It provides quick, affordable, and useful ways to recognize and comprehend germs, enhancing molecular methods and offering crucial information about the taxonomy, ecology, and function of microbes.
13. Molecular identification is a potent tool that improves the precision, dependability, and depth of studies on bacteria and fungi that are endosymbiotic, allowing for their efficient use in a variety of academic and practical contexts.
14. Facilitating growth, isolation, and investigation of microorganisms, culture media development is crucial. It makes it possible for academics and business professionals to take advantage of microbes' advantageous traits for a variety of purposes, from industrial production and environmental management to basic research.
15. Screening endosymbiotic fungi and bacteria for nematicidal activity both in vivo (within a living organism) and in vitro (in a controlled environment outside a living organism) is conducted. A thorough method for identifying and creating safe, long-lasting, and efficient biological control agents for nematode population management is to screen endosymbiotic fungi and bacteria for nematicidal activity. This will support

sustainable agriculture and lessen the environmental impact of chemical nematicides.

16. This thorough approach guarantees a thorough assessment of the nematicidal efficacy of endosymbiotic bacteria and fungus with the goal of creating long-lasting and efficient biocontrol agents against parasitic nematodes.

Acknowledgement

The essential laboratory facilities were provided by the vice chancellor and pro vice chancellor of Eternal University, Baru Sahib, for which the authors are grateful.

References

1. Mitiku M (2018) Plant-parasitic nematodes and their management: a review. Agric Res Technol 8(1):30–38
2. Páez AR (2023) Plant parasitic nematodes and their management. In: Mukherjee S, Ray S (eds) Nematodes – ecology, adaptation and parasitism a focus on new nematicide. Intech Open, pp 1–31
3. Jhamta S, Thakur N, Kaur S (2024) Role of plant parasitic nematodes and dauer larvae against the parasitic adaptation and environmental extremes. In: Exploring innovations in agricultural, biological, chemical, environmental and forensic sciences. Cognizance Publications, pp 74–87
4. Thakur N, Kaur K (2018) Incidence of plant parasitic nematodes associated with capsicum (*Capsicum annuum* L.) in Himachal Pradesh. Indian J Nematol 48:237–239
5. Jhamta S, Thakur N (2023) Occurrence and distribution of *Meloidogyne incognita* infecting vegetable crops in some districts of Himachal Pradesh India. Indian J Nematol 53(1):82–85
6. Kaur S, Thakur N, Jhamta S (2024) Plant parasitic nematodes habitat, diversity and nature of damage. In: Exploring innovations in agricultural, biological, chemical, environmental and forensic sciences, vol 1. Cognizance Publications, pp 88–108
7. Dietel AK, Kaltenpoth M, Kost C (2018) Convergent evolution in intracellular elements: plasmids as model endosymbionts. Trends Microbiol 6(9):755–768
8. Tomar P, Thakur N, Yadav AN (2022) Endosymbiotic microbes from entomopathogenic nematode (EPNs) and their applications as biocontrol agents for agro-environmental sustainability. Egypt J Biol Pest Control 32:80
9. Osman EO, Weinnig AM (2022) Microbiomes and obligate symbiosis of deep-sea animals. Annu Rev Anim Biosci 10(1):151–176
10. Büttner H, Niehs SP, Vandelannoote K, Cseresnyés Z, Dose B, Richter I, Gerst R, Figge MT, Stinear TP, Pidot SJ, Hertweck C (2021) Bacterial endosymbionts protect beneficial soil fungus from nematode attack. Proc Natl Acad Sci 118(37):e2110669118
11. Eleftherianos I, Shokal U, Yadav S, Kenney E, Maldonado T (2016) Insect immunity to entomopathogenic nematodes and their mutualistic bacteria. In: ffrench-Constant RH (ed) The molecular biology of *Photorhabdus* bacteria. Current topics in microbiology and immunology. Springer, Cham, pp 123–156
12. Rosso LC, Lax P, Ciancio A, Colagiero M, Pentimone I (2024) The cryptic microbiota of plant parasitic and entomopathogenic nematodes: diversity, effects on host biology and potential in plant protection. J Pest Sci 26:1–8
13. Wasala SK, Hesse C, Wram CL, Howe DK, Zasada IA, Denver DR (2023) Unraveling microbial endosymbiosis dynamics in plant-parasitic nematodes with a genome skimming strategy. Appl Microbiol 3(4):1229–1248
14. Ayaz M, Zhao JT, Zhao W, Chi YK, Ali Q, Ali F, Khan AR, Yu Q, Yu JW, Wu WC, Qi RD (2024) Biocontrol of plant parasitic nematodes by bacteria and fungi: a multi-omics approach for the exploration of novel nematicides in sustainable agriculture. Front Microbiol 15:1433716
15. Thakur N, Kaur S, Tomar P, Thakur S, Yadav AN (2020) Microbial biopesticides: current

status and advancement for sustainable agriculture and environment. In: Rastegari AA, Yadav AN, Yadav N (eds) New and future developments in microbial biotechnology and bioengineering. Elsevier, pp 243–282

16. Thakur N, Tomar P, Kaur S, Jhamta S, Thakur R, Yadav AN (2021) Entomopathogenic soil microbes for sustainable crop protection. In: Soil microbiomes for sustainable agriculture: functional annotation, pp 529–571
17. Negi R, Sharma B, Kaur S, Kaur T, Khan SS, Kumar S, Ramniwas S, Rustagi S, Singh S, Rai AK, Kour D, Thakur N, Yadav AN (2023) Microbial antagonists: diversity, formulation and applications for management of pest–pathogens. Egypt J Biol Pest Control 33:105
18. Thakur N, Tomar P, Kaur S, Kumari P (2022) Virulence of native entomopathogenic nematodes against major lepidopteran insect species of tomato (*Solanum lycopersicum* L.). J Appl Biol Biotechnol 10(1):1–9
19. Kaya HK, Stock SP (1997) Techniques in insect nematology. In: Lacey LA (ed) Manual of techniques in insect pathology. Academic Press, San Diego, pp 281–324
20. Tomar P, Thakur N, Sidhu AK, Laskar BA, Hashem A, Avila-Quezada GD, Abd_Allah EF (2023) The isolation, identification, and insecticidal activities of indigenous entomopathogenic nematodes (*Steinernema carpocapsae*) and their symbiotic bacteria (*Xenorhabdus nematophila*) against the larvae of *Pieris brassicae*. Horticulturae 9(8): 874
21. Orozco M (2014) A theoretical view of protein dynamics. Chem Soc Rev 43(14):5051–5066
22. Akhurst RJ (1980) Morphological and functional dimorphism in *Xenorhabdus* spp., bacteria symbiotically associated with the insect pathogenic nematodes *Neoaplectana* and *Heterorhabditis*. Microbiology 121(2):303–309
23. Barman S, Chowdhury R, Bhattacharya SS (2022) Screening of endophytes for nematicidal activities. In: Endophytic microbes: isolation, identification, and bioactive potentials. Springer, New York, pp 225–235
24. Chen S, Dickson D (2000) A technique for determining live second-stage juveniles of *Heterodera glycines*. J Nematol 32(1):117

Chapter 24

Screening of Endosymbionts for Biosurfactant Production

Hetvi Naik, Sapna Chandwani, and Natarajan Amaresan

Abstract

This chapter elucidates the methodologies for the screening and characterization of biosurfactants produced by endosymbiotic microorganisms. Endosymbionts are defined as microorganisms residing within the tissues or cells of host organisms and have emerged as potential reservoirs for novel biosurfactants. We delineate cultivation techniques, screening assays, and analytical methods for the identification and characterization of biosurfactants synthesized by these endosymbionts. The methodology encompasses the isolation of endosymbionts from host tissues, preparation of appropriate growth media, and implementation of various screening techniques, including oil displacement assays, emulsification index determination, and surface tension measurements. Furthermore, we discuss the industrial and environmental significance of biosurfactants, emphasizing the potential of endosymbionts as sustainable sources of these amphiphilic compounds.

Key words Biosurfactants, Bacteria, Endosymbionts, Emulsification, Surface tension

1 Introduction

Biosurfactants are amphiphilic molecules synthesized by microorganisms that exhibit surface-active properties and have garnered significant attention in recent years. These compounds have diverse applications in multiple industries, including pharmaceuticals, cosmetics, food processing, and environmental remediation [1]. In comparison to their synthetic counterparts, biosurfactants offer several advantages such as reduced toxicity, enhanced biodegradability, and improved environmental compatibility [2].

Endosymbionts, microorganisms that have evolved to reside within the tissues or cells of plants or animals, represent an underexplored source of novel bioactive compounds, including biosurfactants. These microorganisms have developed unique metabolic capabilities to adapt to their host environments, potentially leading to the production of structurally diverse and functionally distinct biosurfactants [3].

N. Amaresan and Sapna Chandwani (eds.), *Endosymbionts*, Springer Protocols Handbooks,
https://doi.org/10.1007/978-1-0716-4771-4_24,

This chapter focuses on the methodological approaches for screening endosymbionts for biosurfactant production. Here, we present techniques for isolating endosymbionts from various host organisms, preparing appropriate growth media, and employing a range of screening assays to detect and characterize biosurfactant production. The methods described herein will enable researchers to explore the potential of endosymbionts as a source of novel biosurfactants for various industrial and environmental applications.

2 Materials

2.1 Equipment

1. Class II Biological Safety Cabinet.
2. Incubator (temperature range: 25–37 °C, ±0.1 °C accuracy).
3. Refrigerated centrifuge (capable of 10,000× *g*).
4. UV–Vis spectrophotometer.
5. Du Noüy ring tensiometer.
6. Phase contrast microscope.
7. Autoclave (121 °C, 15 psi).
8. pH meter (accuracy ±0.01 pH units).
9. Analytical balance (precision: 0.1 mg).
10. Vortex mixer.
11. Orbital shaking incubator (25–250 rpm).

2.2 Glassware and Plasticware

1. Sterile Petri dishes (90 mm diameter).
2. Erlenmeyer flasks (250 mL, 500 mL), borosilicate glass.
3. Sterile test tubes (10 mL, 20 mL).
4. Microscope slides and coverslips.
5. Micropipettes (P20, P200, P1000) and sterile tips.
6. Sterile centrifuge tubes (15 mL, 50 mL).
7. Borosilicate glass beakers and graduated cylinders.

2.3 Chemicals and Reagents

1. *Nutrient agar*

   Suspend 23 g of nutrient agar powder in 1 L of deionized water. Heat to boiling with constant stirring to dissolve completely. Autoclave at 121 °C for 15 min and cool to approximately 50 °C before aseptically dispensing into sterile Petri dishes.

2. *Nutrient broth*

   Dissolve 13 g of nutrient broth powder in 1 L of deionized water. Mix thoroughly and distribute into appropriate containers. Autoclave at 121 °C for 15 min.

3. *Mineral salt medium components*

NaCl	10 g
KH_2PO_4	0.5 g
K_2HPO_4	1.0 g
$MgSO_4.7H_2O$	0.5 g
KCl	0.5 g
$CaCl_2$	0.01 g
Yeast extract	0.1 g
Glucose	20.0 g
$FeSO_4{\cdot}7H_2O$	0.5 g
$MnSO_4{\cdot}H_2O$	0.5 g
H_3BO_3	0.1 g
$CuSO_4{\cdot}5H_2O$	0.01 g
$ZnSO_4{\cdot}7H_2O$	0.1 g
$Na_2MoO_4{\cdot}2H_2O$	0.1 g
$CoCl_2{\cdot}6H_2O$	0.1 g

Dissolve all components except glucose in deionized water. Adjust pH to 7.0 ± 0.1 using 1 M NaOH or 1 M HCl. Prepare glucose solution separately and sterilize by membrane filtration (0.22 μm). Autoclave the base medium at 121 °C for 15 min. After cooling to room temperature, aseptically add the sterile glucose solution and filter-sterilized trace element solution.

4. *Phosphate- buffer saline (PBS)*

NaCl	8 g
KCl	0.2 g
Na_2HPO_4	1.44 g
KH_2PO_4	0.24 g
Distilled water	1000 mL
pH	7.4

Dissolve the above-mentioned quantities of NaCl, KCl, Na_2HPO_4, and KH_2PO_4 in 800 mL of deionized water. Adjust the pH to 7.4 using 1 M HCl or 1 M NaOH as needed. Add deionized water to bring the final volume to 1 liter. Sterilize the solution by autoclaving at 121 °C for 15 min or by filter sterilization using a 0.22-μm membrane filter. Store at room temperature or 4 °C. Discard if there is any sign of contamination or precipitation (*see* **Note 1**).

5. Vegetable oil (e.g., extra virgin olive oil, sunflower oil).
6. Methylene blue (≥82% purity).
7. Cetyltrimethylammonium bromide (CTAB).

3 Methods

3.1 Isolation of Endosymbionts

3.1.1 Surface Sterilization of Host Tissue

1. Wash the plant or animal tissue thoroughly with sterile deionized water and immerse in 70% (v/v) ethanol for 60 s.
2. Transfer to 2% (w/v) sodium hypochlorite solution for 180 s. Rinse three times with sterile deionized water.

3.1.2 Isolation of Endosymbionts

1. Aseptically macerate the surface-sterilized tissue in sterile phosphate-buffered saline (PBS, pH 7.4) and prepare serial dilutions (10^{-1} to 10^{-6}) of the macerate using sterile PBS.
2. Inoculate 100 μL of each dilution onto nutrient agar plates using the spread plate technique and incubate at 28 ± 1 °C for 24–72 h.
3. Select morphologically distinct colonies for further screening and purification (*See* **Notes 2, 3**).

3.2 Screening for Biosurfactant Production

3.2.1 Oil Spreading Assay

1. Add 50 mL of deionized water to a 150-mm-diameter Petri dish and carefully pipette 20 μL of crude oil onto the water surface to form a thin film.
2. Gently add 10 μL of cell-free culture supernatant to the center of the oil film.
3. Measure the diameter of the clear zone formed after 60 s.
4. Calculate the oil displacement area using the formula:

   $A = \pi(r^2)$, where r is the radius of the clear zone [4].

3.2.2 Emulsification Index (E24)

1. In a clean test tube, combine 2 mL of cell-free culture supernatant with 2 mL of vegetable oil (*see* **Note 4**).
2. Vortex at maximum speed for 120 s to ensure thorough mixing.
3. Allow the mixture to stand undisturbed at room temperature for 24 h.
4. Calculate the E24 index using the formula:

   E24 = (Height of emulsion layer/Total height of the liquid column) × 100 [5]

3.2.3 Surface Tension Measurement

1. Measure the surface tension of cell-free culture supernatant using a Du Noüy ring tensiometer at 25 °C.
2. Use deionized water as a control (surface tension approximately 72 mN/m at 25 °C).
3. A significant reduction in surface tension (typically below 40 mN/m) indicates biosurfactant production [6].

3.2.4 Blue Agar Plate Method

1. Prepare mineral salt agar supplemented with 0.2 g/L cetyltrimethylammonium bromide (CTAB) and 0.005 g/L methylene blue.
2. Inoculate the plates with endosymbiont isolates using a sterile inoculation loop.
3. Incubate at 28 ± 1 °C for 48–72 h.
4. Observe the formation of dark blue halos around colonies, indicating the production of anionic biosurfactants (*See* **Note 5**) [7].

4 Notes

1. For convenience, you can prepare a 10× stock solution of PBS and dilute it to 1× as needed. However, always check and adjust the pH of the 1× solution before use.
2. Maintain strict aseptic conditions during all stages of endosymbiont isolation and cultivation to prevent contamination.
3. Optimize growth conditions (temperature, pH, carbon source) for each isolate to maximize biosurfactant production. Consider using a design of experiment (DoE) approach for optimization.
4. When inducing biosurfactant production, consider using alternative carbon sources such as vegetable oils or hydrocarbons, as some strains may preferentially produce biosurfactants on hydrophobic substrates.
5. Perform all screening assays in triplicate to ensure statistical validity and reproducibility. Report results as mean ± standard deviation.

References

1. Banat IM, Franzetti A, Gandolfi I, Bestetti G, Martinotti MG, Fracchia L, Smyth TJ, Marchant R (2010) Microbial biosurfactants production, applications and future potential. Appl Microbiol Biotechnol 87:427–444
2. Marchant R, Banat IM (2012) Microbial biosurfactants: challenges and opportunities for future exploitation. Trends Biotechnol 30(11): 558–565
3. Fardami AY, Kawo AH, Yahaya S, Lawal I, Abubakar AS, Maiyadi KA (2022) A review on biosurfactant properties, production and producing microorganisms. J Biochem Microbiol Biotechnol 10(1):5–12

4. Morikawa M, Hirata Y, Imanaka T (2000) A study on the structure–function relationship of lipopeptide biosurfactants. Biochim Biophys Acta Mol Cell Biol Lipids 1488(3):211–218
5. Cooper DG, Goldenberg BG (1987) Surface-active agents from two *Bacillus* species. Appl Environ Microbiol 53(2):224–229
6. Youssef NH, Duncan KE, Nagle DP, Savage KN, Knapp RM, McInerney MJ (2004) Comparison of methods to detect biosurfactant production by diverse microorganisms. J Microbiol Methods 56(3):339–347
7. Siegmund I, Wagner F (1991) New method for detecting rhamnolipids excreted by pseudomonas species during growth on mineral agar. Biotechnol Tech 5(4):265–268

Chapter 25

Screening of Endosymbionts as Anthelmintic Agent

Sapna Chandwani, Hetvi Naik, and Natarajan Amaresan

Abstract

In spite of several studies on endosymbionts, the research on the potential of endosymbionts as anthelmintic agents has still not yet been fully exploited. This innovative approach of endosymbionts to control parasitic infections helps understand the endosymbionts' interactions with the helminths/parasites and resolves issues such as drug resistance of helminthic infections. Screening of endosymbionts as anthelmintic agents has attracted interest in developing novel therapy which could target the neglected tropical diseases and infections caused due to helminths. Not only this, the current area of research also plays a key role in understanding the various aspects of symbiotic microorganisms in parasite biology.

Key words Endosymbionts, Anthelmintic, Bioactive compounds, Parasitic infections

1 Introduction

Parasites such as trematodes, nematodes, tapeworms, roundworms, and cestodes pose a major threat to crop plants as well as to livestock in terms of food security and economy. To control parasitic helminths, several conventional anthelmintic treatments are available, but they face challenges such as safety concerns, drug resistance, and environmental impacts. One of the research studies reported the use of conventional anthelmintic drugs such as benzimidazoles and ivermectin to control the parasitic worms in livestock such as cattle, sheep, and goats. However, a major issue was addressed when the parasites were found to be resistant to the anthelmintics, thereby creating an urgent need for an alternative approach to controlling the parasitic infections [1]. To address such challenges, the use of endosymbionts is raised as a promising targeted approach, offering a novel and eco-friendly solution [1].

To control parasitic infections, endosymbionts have immense potential by targeting symbionts such as bacterial or fungal symbionts which rely on the parasitic worms for survival. Compared to conventional anthelmintic drugs which directly try to eradicate the parasite, a less harmful approach of endosymbiont use indirectly

N. Amaresan and Sapna Chandwani (eds.), *Endosymbionts*, Springer Protocols Handbooks,
https://doi.org/10.1007/978-1-0716-4771-4_25,

reduces the ability of the helminths to develop resistance against anthelmintics and easily manages the parasitic infections in livestock [2]. Endosymbiont-based treatment of helminths minimizes the ecological damage by targeting the biological system within the parasite only, while other conventional anthelmintics affect the environment particularly soil and aquatic systems by leaving residues in the environment which indirectly impacts the nontarget species and ecosystems [3]. Moreover, endosymbiont-based treatment of parasitic loads is often a cost-effective approach for the farmers to attain sustainable agriculture; for example., symbionts like *Wolbachia* treated by antibiotics (endosymbiont-targeted therapy) are found to be longer-lasting and require only fewer and low doses, whereas conventional anthelmintic treatments are costly as they require repeated dosage applications and are not long-lasting [4]. This book chapter describes how endosymbionts can be screened out to be used as an anthelmintic agent.

2 Materials

1. Tissues of host organisms (plants, insects, or marine animals) from which endosymbionts can be isolated.
2. Light microscope/inverted microscope.
3. Glasswares such as Petri dishes, pipette, glass bottles, flasks, and micropipette.
4. *Media:* Nutrient agar medium, nutrient broth, Luria–Bertini broth, tryptic soy agar, yeast peptone dextrose medium, and specific growth media.
5. *Stains:* Trypan blue or 3-(4,5-dimethylthiazolyl-2)-2,5-diphenyltetrazolium bromide (MTT).
6. *Instruments:* Centrifuge, sonicator, vortex, RT-PCR, and mass spectrometer.
7. Helminths as model organisms for testing.
8. Animal model such as mice, rats, sheep, or goats for in vivo screening of the endosymbiont extract.

3 Methods

3.1 Sample Collection and Isolation of Endosymbionts

1. Collect the tissue samples from the host organisms such as nematodes, crustaceans, plants, marine animals, or insects using sterile equipment in order to avoid contamination to isolate the endosymbiotic microbes (bacterial, fungal, or protozoan) (*see* **Note 1**).

2. Isolate the endosymbionts from the host tissue by culturing these microorganisms on a selective media such as brain heart infusion agar, marine broth, marine agar, nutrient agar, tryptic soy agar, yeast peptone dextrose medium, LB medium, and artificial pond water (*see* **Note 2**).
3. Incubate the endosymbionts under optimal conditions such as oxygen requirement, temperature, and incubation time in order to attain the appropriate growth of these organisms [5].
4. Characterize and identify the endosymbionts by 16S rRNA gene sequencing.

3.2 Extraction of Metabolites (Bioactive Compounds)

5. Lyse the cells using methods such as chemical lysis, enzymatic digestion, sonication, solvent extraction, or freeze-thaw cycles to release the intracellular compounds from the endosymbionts.
6. Harvest the cells from the endosymbiotic microbes by centrifugation at 8000–10000 rpm for 15 min at 37 °C.
7. Filtrate the suspension to remove the cell debris and make sure that the prepared endosymbiotic extract contains the extracted metabolites [6].

3.3 Screening of Endosymbiont Extract Against Helminths

8. For testing the efficacy of the prepared extract, select the appropriate helminths as a model organism such as *Caenorhabditis elegans*, *Brugia malayi*, *Haemonchus contortus*, or *Schistosoma mansoni* and culture them under appropriate conditions of temperature, humidity, and nutrient availability (*see* **Note 3**).
9. Now, treat the cultured helminths (larvae or adults' stage) with different concentrations of the endosymbiont extracts for different time intervals of 24–72 h to assess the effect of the prepared extract on worm viability, motility, or development.
10. Observe the activity of the parasitic worms for a set period of 24–72 h.
11. The reduced movement, paralyzed effect, or death of the worms indicates the efficacy of the endosymbiont extract as an anthelmintic agent.
12. The viability of the worms is then assessed by using MTT or trypan blue stains to differentiate live worms and dead worms.
13. Determine the minimum inhibitory concentration (MIC) or lethal dosage (LD50) of the extract required to inhibit or kill the helminths at different concentrations.
14. Perform real-time PCR (RT-PCR) or RNA-seq to analyze the gene expression of helminths targeted by the prepared endosymbiont extract [7].

3.4 In Vivo Screening of the Endosymbiont Extract

15. After performing the in vitro studies, analyze the efficacy of the endosymbiont-derived extract in an animal model such as mice, rats, or sheep.
16. Infect the animal model with appropriate parasitic worms (helminths). For example, mice or rats can be infected with the helminths *Trichuris muris*, *Schistosoma mansoni*, or *Heligmosomoides polygyrus* (*see* **Note 4**).
17. Monitor the animal on a regular basis till it gets infected with the parasite (*see* **Note 5**).
18. Treat the infected animal with the prepared extract either orally, topically, or via injection (*see* **Note 6**).
19. To determine the effective dose of the extract required, treat the helminth-infected animal with multiple doses of the endosymbiont extract such as 50 mg kg^{-1} and 100 mg kg^{-1} [8].
20. After treatment with the endosymbiont extract, measure a reduction in the number of adult worms, and in the case of fecal-dwelling helminths, measure the number of eggs per gram (egg counts) between the treated and untreated animals [9].
21. Observe and evaluate the toxicity symptoms in the animals such as behavior, body weight, or biochemical markers (liver enzymes, proteases) to ensure the safety of the prepared endosymbiont extract (*see* **Note 7**) [10].
22. Examine the tissue samples of the treated animals histologically to observe whether there occurred any damage or toxicity.
23. To identify the active compounds of the metabolites of the endosymbiont extract, perform HPLC, LC–MS, or GC–MS analysis.

4 Notes

1. Selection of host organisms should be appropriate that can harbor endosymbionts.
2. Specialized growth media with host-specific supplements is required as all the endosymbionts cannot be cultured in an easy manner under in vitro conditions without their host.
3. Before testing the endosymbiont extract against helminths, make sure that the prepared extract should be filtered and concentrated.
4. Infect the animal model with a controlled number of helminth eggs or larvae.
5. Depending on the life cycle of the parasite, infection may occur; generally, it may take 7–14 days.

6. Based on the formulation of the extract and the parasite's location, choose the administration route of the prepared extract.
7. To examine any alterations in the biochemical markers of animals after treatment with the extract, perform the biochemical analysis to understand the mechanism responsible for causing the death of the worms (helminths).

References

1. Kaplan RM, Vidyashankar AN (2012) An inconvenient truth: global warming and anthelmintic resistance. Vet Parasitol 186:70–78
2. Xie S, Lan Y, Sun C, Shao Y (2019) Insect microbial symbionts as a novel source for biotechnology. World J Microbiol Biotechnol 35: 1–7
3. Vokřál I, Podlipná R, Matoušková P, Skálová L (2023) Anthelmintics in the environment: their occurrence, fate, and toxicity to non-target organisms. Chemosphere 345: 140446
4. Strydom T, Lavan RP, Torres S, Heaney K (2023) The economic impact of parasitism from nematodes, trematodes and ticks on beef cattle production. Animals 13:1599
5. Lefoulon E, Foster JM, Truchon A, Carlow CK, Slatko BE (2020) The Wolbachia symbiont: here, there and everywhere. In: Symbiosis: cellular, molecular, medical and evolutionary aspects, vol 69, pp 423–451
6. Singh M, Kumar A, Singh R, Pandey KD (2017) Endophytic bacteria: a new source of bioactive compounds. 3 Biotech 7:1–4
7. Burns AR, Luciani GM, Musso G, Bagg R, Yeo M, Zhang Y, Rajendran L, Glavin J, Hunter R, Redman E, Stasiuk S (2015) *Caenorhabditis elegans* is a useful model for anthelmintic discovery. Nat Commun 6:7485
8. Sepúlveda-Crespo D, Reguera RM, Rojo-Vázquez F, Balaña-Fouce R, Martínez-Valladares M (2020) Drug discovery technologies: *Caenorhabditis elegans* as a model for anthelmintic therapeutics. Med Res Rev 40:1715–1753
9. Boelow H, Krücken J, Thomas E, Mirams G, von Samson-Himmelstjerna G (2022) Comparison of FECPAKG2, a modified mini-FLOTAC technique and combined sedimentation and flotation for the coproscopic examination of helminth eggs in horses. Parasit Vectors 15:166
10. Upadhyay RK (2010) Essential oils: antimicrobial, antihelminthic, antiviral, anticancer and antiinsect properties. J Appl Biosci 36:1–22

Chapter 26

Screening of Endosymbionts for Antibiotic and Probiotic Activity

Riuh Wardhani and Dirayah Rauf Husain

Abstract

Screening endosymbionts for antibiotic and probiotic activity involves evaluating their efficacy in inhibiting pathogenic microorganisms and promoting beneficial microbial growth. This process utilizes various bioassays such as diffusion and dilution methods, thin-layer chromatography (TLC)–bioautography, and time–kill tests to assess endosymbionts' antibiotic properties. Additionally, screening for probiotic activity considers factors such as acid and bile salt tolerance, adherence to intestinal surfaces, and antibiotic activity against pathogens. This study discusses antibiotic screening methodologies, such as agar diffusion and microdilution, and probiotic screening techniques, such as acid tolerance, bile salt tolerance, and gastrointestinal simulation.

Key words Endosymbionts, Antibiotic activity, Probiotic activity

1 Introduction

Endosymbionts, microorganisms living within a host organism, hold promise as sources of novel antibiotics and probiotics. They are significant because they produce secondary metabolites with potential pharmaceutical, industrial, and agricultural value. These microorganisms establish symbiotic relationships with their hosts and contribute to various health benefits.

The potential of endosymbionts as antibiotics is based on their ability to produce bioactive compounds that inhibit the growth of pathogenic microorganisms. The principal mechanism of these antibiotic compounds involves disrupting essential processes in target pathogens, thereby preventing their growth and proliferation.

As probiotics, endosymbionts must meet specific criteria to confer health benefits to the host. These criteria include the ability to survive the harsh conditions of the gastrointestinal environment, adhere to intestinal cells, and inhibit pathogenic bacteria. Probiotic

N. Amaresan and Sapna Chandwani (eds.), *Endosymbionts*, Springer Protocols Handbooks,
https://doi.org/10.1007/978-1-0716-4771-4_26,

endosymbionts play a crucial role in maintaining the host's microecological balance, improving intestinal function, and stimulating digestion and immune responses.

The screening of endosymbionts for antibiotic potential involves various methods [1–4].

1. *Diffusion Assays*: These include disk diffusion and well diffusion methods. In disk diffusion, antibiotic-producing microorganisms are placed on a nutrient agar plate inoculated with the target pathogen. The inhibition of pathogen growth around the disk indicates antibiotic activity. Well diffusion follows a similar principle, where wells are punched in the agar, filled with microbial extracts, and the zone of inhibition is measured.
2. *Dilution Techniques*: Broth and agar dilution methods are used to determine the minimum inhibitory concentration (MIC) of the antibiotic compounds. In broth dilution, the microorganism is cultured in liquid media containing different concentrations of the antibiotic agent. Agar dilution involves incorporating the antibiotic agent into agar plates at varying concentrations to assess microbial growth.
3. *TLC–Bioautography*: Thin-layer chromatography (TLC) combined with bioautography is used to separate and identify bioactive compounds. After separation by TLC, the plate is overlaid with a microbial suspension, and zones of inhibition on the chromatogram indicate the presence of antibiotic compounds.
4. *Time–Kill Tests*: These tests measure the bactericidal effects of antibiotic agents over time. The microorganism is exposed to the agent, and samples are taken at various time points to determine the reduction in viable cell counts.

For probiotics, the screening methods focus on evaluating the following properties [5]:

1. *Tolerance to Acidic and Bile Conditions*: Probiotic strains must survive the acidic pH of the stomach and the presence of bile in the intestines. Tests involve exposing the microorganisms to low pH and bile salts and assessing their viability.
2. *Adherence to Intestinal Cells*: The ability of probiotic strains to adhere to intestinal epithelial cells is crucial for colonization and exerting their beneficial effects. Adhesion assays involve incubating the probiotic strains with intestinal cell lines and measuring the extent of adherence.
3. *Inhibition of Pathogenic Bacteria*: Probiotic strains are screened for their ability to inhibit the growth of pathogenic bacteria. This can be assessed using co-culture techniques where

probiotics and pathogens are cultured together, and the inhibition of pathogen growth is measured.

These screening methods provide comprehensive insights into the beneficial properties of endosymbionts, enabling the identification and development of new antibiotics and probiotics with significant health benefits. In this study, we focus on the specific methodologies used for antibiotic screening, namely, the agar diffusion and microdilution methods. Additionally, we provide an overview of three prevalent techniques for probiotics screening: acid tolerance, bile salt tolerance, and gastrointestinal simulation.

2 Materials

2.1 Media

Mueller–Hinton agar (MHA), Mueller–Hinton broth (MHB), MRS broth: de Man, Rogosa, and Sharpe medium (MRSA) and tryptic soy agar (TSA).

2.2 Solutions and Single-Use Materials

Distilled/purified water, hydrochloric acid (HCl), sodium bicarbonate ($NaHCO_3$), phosphate-buffered saline (PBS), 0.9% NaCl solution, ciprofloxacin, bile salts, pepsin, trypsin, blank disk, ethyl acetate, and resazurin dye.

2.3 Equipment

Spectrophotometer, centrifuge, sonicator, incubator, shaker, autoclave Petri dish, cork borer, 96-well microtitration plates, vernier caliper, and micropipette.

3 Methods

Prepare all media and solutions using purified or distilled water and analytical-grade reagents. Sterilize all media at 121 °C for 20 min at 1 atm before use. Sterilize Petri dishes and cork borers at 160–170 ° C for 2–3 h. Prepare and store all reagents at room temperature unless otherwise indicated. Ensure complisance with all waste disposal regulations when disposing of waste materials.

3.1 Antibiotic Screening: Diffusion Methods (Agar Well and Disk Diffusion)

1. Pre-culture of Endosymbiont Bacteria Isolates

Transfer 1–4 mL of endosymbiont bacterial isolates from TSA to 50 mL of MHB. Incubate the cultures in a shaker at 120 rpm for 4 days (4 × 24 h) (*see* **Note 1**).

2. Preparation of Test Bacteria

Inoculate pure cultures of test bacteria (e.g., *Salmonella typhi* and *Staphylococcus aureus*) onto slanted Muller–Hinton agar (MHA) and incubate at 37 °C for 24 h. Suspend the 24-h-old cultures in a sterile 0.9% NaCl solution to obtain a bacterial

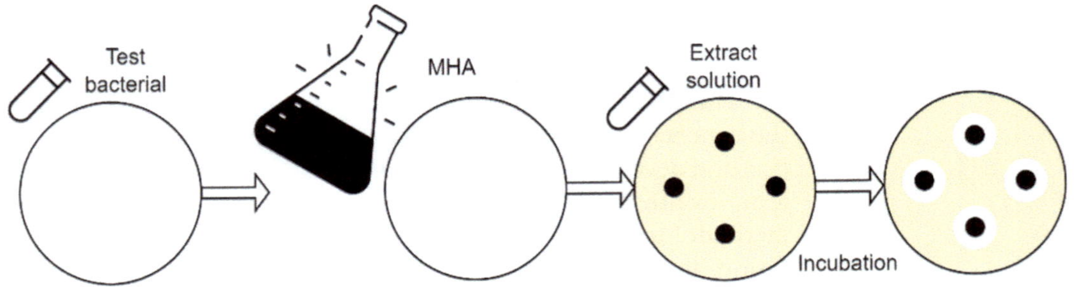

Fig. 1 Schematic of agar well diffusion method

suspension. Adjust the density of the suspension using a spectrophotometer at 580 nm to achieve a 25% transmittance, which corresponds to a concentration of 10^{-8} CFU/mL (*see* **Note 2**).

3. Preparation of Control Solution (Positive Control)

Prepare a ciprofloxacin solution at a concentration of 30 ppm by dissolving 0.003 grams of ciprofloxacin in 100 mL of distilled water (*see* **Note 3**).

4. Antibiotic Testing Procedure

Centrifuge the endosymbiont culture to separate it into supernatant and pellet. Extract intracellular compounds by sonicating the pellet. For extracellular compounds, use the supernatant directly (*see* **Note 4**).

5. Agar Well Diffusion Assay

Pour 1 mL of the test bacterial suspension into sterile Petri dishes. Add MHA (50–75 mL) and mix thoroughly before allowing it to solidify. Then, a hole with a diameter of 6–8 mm is punched aseptically with a sterile cork borer or a tip, and a volume (20–100 mL) of the extract solution and positive control at the desired concentration is introduced into the well. Incubate the plates at 37 °C for 24 h, then measure the diameter of the inhibition zones using a vernier caliper (*see* **Note 5**) (Fig. 1).

6. Disk Diffusion Assay

Soak sterile blank disks (10 mm diameter) in the old bacterial culture and positive control solution for 15 min. Place the soaked disks aseptically on the surface of the inoculated MHA plates using sterile tweezers, ensuring ±2 cm between disks and from the edge of the plate.

3.2 Antibiotic Screening: Dilution Methods (Broth Microdilution)

1. Follow **steps 1**, **2**, and **3** in Subheading 3.1. Diffusion methods
2. Extraction of Antibiotic Compounds from Endosymbiont

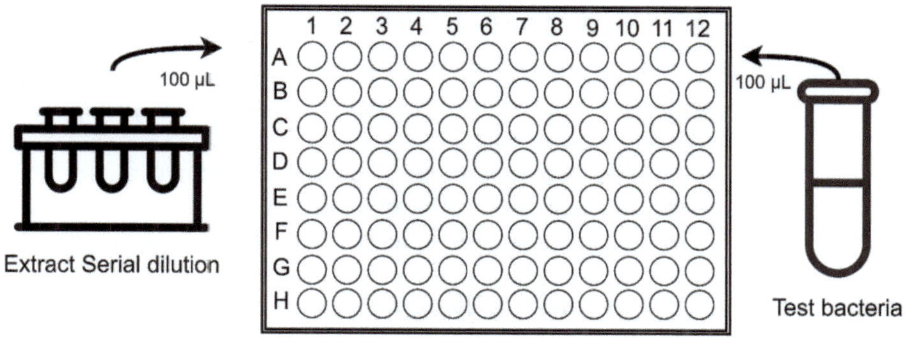

Fig. 2 Schematic of microdilution method

The extraction of antibiotic compounds from endosymbionts was carried out using an adapted protocol. Initially, the culture filtrate of the endosymbiont was mixed with an equal volume of ethyl acetate and subjected to overnight agitation to facilitate the interaction between the filtrate and the solvent. Subsequently, phase separation was performed using a separating funnel, and the bioactive metabolites were collected from the upper layer. The collected layer was then concentrated using a vacuum rotary evaporator to remove ethyl acetate and isolate the bioactive compounds (*see* **Note 6**).

3. Broth Microdilution Procedure

Serial dilutions of the antibiotic agent (endosymbiont extracts) were prepared in Mueller–Hinton broth (MHB), typically ranging from 1 mg/mL to 32 mg/mL (*see* **Note 7**). A volume of 100 μL of each dilution was dispensed into the wells of a 96-well microtitration plate. Subsequently, 100 μL of bacterial inoculum (5×10^5 CFU/mL) was added to each well, ensuring a final volume of 200 μL per well. The microtitration plates were incubated at 37 ° C for 16–20 h without shaking. The minimum inhibitory concentration (MIC) was defined as the lowest concentration of the antibiotic agent that completely inhibited visible bacterial growth. This was assessed either visually or with the aid of a spectrophotometer (Fig. 2).

4. Endpoint Determination

Visual Inspection The MIC is the well with the lowest concentration of antibiotic agent showing no visible growth.

Colorimetric Assay Use resazurin dye as a growth indicator. Add 30 μL of 0.02% resazurin solution to each well and incubate for 2–4 h. Wells that remain blue indicate no bacterial growth, confirming the MIC.

5. Calculation of Minimum Bactericidal Concentration (MBC)

Subculturing: From each well showing no visible growth, subculture 10 μL onto MHA plates. Incubate the plates at 37 °C for 24 h (*see* **Note 8**).

MBC determination: The MBC is defined as the lowest concentration of the antibiotic agent at which no bacterial colonies grow on the agar plates, indicating 99.9% bacterial killing.

3.3 Probiotic Screening

1. *Preparation of Bacterial Isolates*

Cultivate endosymbionts isolate on 50 mL MRSB. Incubate the cultures in a shaker at 120 rpm for 1–2 × 24 h.

2. *Acid Tolerance Test*

Prepare three test tubes that contain 5–10 mL MRSB. Adjust the pH of MRS broth to 2.0, 3.0, and 6.4 (control) using 0.1 N HCl. Then, inoculate 10 μL of 24-h bacterial culture into 1 mL of each pH-adjusted MRS broth. Incubate at 37 °C for 2 h. Use the plate count method to determine the number of viable bacteria. Calculate the survival rate as log values of colony-forming units per milliliter (CFU/mL) (*see* **Note 9**).

A high log value of CFU/mL after incubation indicates that the bacterial isolate can withstand acidic conditions. This suggests good potential for the isolate to survive the acidic environment of the stomach, which is essential for a probiotic candidate. A significant drop in the log value indicates that the bacterial isolate is not acid-tolerant. Such isolates may not survive the gastric environment, reducing their effectiveness as probiotics.

3. *Bile Salt Tolerance Test*

Prepare two test tubes that contain 5–10 mL MRSB. Add 0.3% (w/v) bile salts to MRS broth, then inoculate 1% overnight bacterial cultures into MRS broth with and without bile salts. Incubate at 37 °C and monitor growth by measuring absorbance at 600 nm. Determine the bacterial growth due to bile salt inhibition by comparing the absorbance increase in test and control cultures.

If the absorbance shows a significant increase in both the test and control cultures, this suggests that the isolate can tolerate bile salts. A noticeable decrease in absorbance for the test culture compared to the control suggests that the bacterial isolate is sensitive to bile salts.

4. *Simulation of Gastrointestinal Tract Tolerance*

3.3.1 Preparation of Simulated Gastric and Intestinal Fluids

Gastric Fluid Dissolve 0.35 g pepsin in 100 mL of 0.2% sterile NaCl solution, and adjust pH to 2.5.

Intestinal Fluid Dissolve 0.1 g trypsin, 1.8 g bile salts, 1.1 g $NaHCO_3$, and 0.2 g NaCl in 100 mL distilled water, and adjust pH to 8.0.

Inoculation and Incubation Centrifuge overnight bacterial cultures and resuspend in PBS at pH 7.0. Transfer 10% of the bacterial suspension into simulated gastric and intestinal fluids. Incubate gastric samples for 3 h and intestinal samples for 4 h at 37 °C.

Viability Assessment Count viable bacteria at 0 and 3 h (gastric) and 0 and 4 h (intestinal) using the plate count method. Calculate survival rates.

If the viable bacterial count remains high after incubation in both simulated gastric (for 3 h) and intestinal fluids (for 4 h), this indicates a strong tolerance to gastrointestinal conditions. The survival of bacteria in both environments suggests robust potential as a probiotic. A significant reduction in the viable bacterial count in either the gastric or intestinal fluid indicates poor tolerance. This suggests that the isolate may not survive the passage through the gastrointestinal tract, reducing its efficacy as a probiotic.

4 Notes

1. Pre-culture Periods: Pre-culture periods should be determined based on the bacterial growth curve. High antibiotic production typically occurs at the end of the exponential phase and the beginning of the stationary phase.
2. Bacterial Test Suspension Concentration: The concentration of the bacterial test suspension can be measured using spectrophotometry at 580 nm with 25% transmittance or using the McFarland standard. While the McFarland standard may be necessary, spectrophotometry can be used under certain conditions.
3. Positive Control Selection: When selecting a positive control, consider the effectiveness of synthetic antibiotics. In this study, we used ciprofloxacin due to its effectiveness against both Gram-positive and Gram-negative bacteria.
4. Extraction of Supernatants or Pellets: For endosymbionts with low antibiotic activity, the extraction of supernatants or pellets may be required. Suitable solvents should be selected for the target compound, and evaporation should be performed after extraction to prevent any inhibitory effects caused by residual solvents.

5. Inhibition Zone Analysis: Record and analyze inhibition zone diameters to determine the antibacterial activity of endosymbiont extracts. The results can be interpreted based on the size of the inhibition zones formed.
6. Modification of Antibiotic Compound Extraction: Various modifications can be employed for antibiotic compound extraction, such as freeze-drying or spray-drying before extraction, using different solvents, and adjusting solvent ratios. Selecting the appropriate solvent is crucial and should be based on the specific target compound to be extracted.
7. Dilution Accuracy: Perform dilutions carefully to ensure accuracy. Serial dilutions can be done in tube tests or microplates. Use calibrated pipettes and sterile techniques to prevent cross-contamination.
8. Subculturing Techniques: Subculture carefully to avoid transferring nontarget microorganisms. Accurate determination of the minimum bactericidal concentration (MBC) is crucial for evaluating bactericidal activity.
9. Plate Count Method: The plate count method provides comprehensive information regarding bacterial survival rates. For probiotic pre-screening, visual observation of bacterial growth in the tube is often sufficient.
10. Spectrophotometry Limitations: Limitations in spectrophotometry access can be addressed by visual observation in test tubes. Turbidity indicates microbial growth, which signifies positive results for the tested parameters.

References

1. Balouiri M, Sadiki M, Ibnsouda SK (2016) Methods for in vitro evaluating antibiotic activity: a review. J Pharm Anal 6(2):71–79
2. $CLSI_a$ (2012) Performance standards for antibiotic disk susceptibility tests. Approved Standard, 7th ed, CLSI document M02-A11
3. $CLSI_b$ (2012) Methods for dilution antibiotic susceptibility tests for bacteria that grow aerobically. Approved Standard, 9th ed, CLSI document M07-A9
4. Marston A (2011) Thin-layer chromatography with biological detection in phytochemistry. J Chromatogr A 1218(19):2676–2683
5. Zhang B, Wang Y, Tan Z, Li Z, Jiao Z, Huang Q (2016) Screening of probiotic activities of lactobacilli strains isolated from traditional Tibetan Qula, a raw yak Milk cheese. Asian Australas J Anim Sci 29(10):1490–1499. https://doi.org/10.5713/ajas.15.0849

Part IV

Methods to Detect Bacterial Infection of Endosymbionts in Insects

Chapter 27

Diagnostic PCR to Detect Infection of Bacterial Endosymbionts in Insects

Shraddha Saha

Abstract

Many arthropods (insects) have been associated with bacterial endosymbionts. These bacterial endosymbionts provide various advantages to insects, including nutritional, physiological, environmental, and evolutionary benefits. This is because of the various advantages of endosymbionts in their insect hosts. This has caused serious global issues leading to crop damage worldwide. This chapter describes the procedure of polymerase chain reaction (PCR) to detect such bacterial endosymbionts in insects.

Key words Endosymbionts, Crops, PCR, Amplification

1 Introduction

Symbiosis refers to the permanent association between two or more distinct individuals. The organism that lives inside the cell of another is called an endosymbiont. It has been estimated that around 15% of all insects possess bacterial endosymbionts association [1, 2]. Insects belonging to psyllids, aphids, mealybugs, and white flies have been reported to have bacterial endosymbiont associations. Insect endosymbionts are categorized into primary and secondary endosymbionts. The former has an obligatory relationship with the insect host, which provides essential amino acids and shows phylogenetic similarity with their host [3, 4]. Secondary endosymbionts have a facultative relationship and evolutionary history with their host [4, 5].

Various insects are sap-sucking and feed on the phloem sap of a wide range of crops such as brinjal, cabbage, radish, tomato, beans, cucurbits, and potato. Additinally they damage cash crops like cotton, sunflower, tobacco and legumes, directly or indirectly by sucking phloem sap. A few crops also harbor both primary and secondary endosymbionts. These endosymbionts are responsible

N. Amaresan and Sapna Chandwani (eds.), *Endosymbionts*, Springer Protocols Handbooks,
https://doi.org/10.1007/978-1-0716-4771-4_27,

for conferring important abilities to their hosts and are also known to contribute to virus transmission [6].

These endosymbiotic microorganisms generally remain uncultured because they are highly adapted to the internal environment of their host organisms and cannot survive outside the host [7]. Therefore, PCR analysis of endosymbiont microbial genes becomes a powerful approach to infer the systematic affinities of these endosymbionts [8–15].

2 Materials

2.1 DNA Extraction

DNA is extracted from the insect samples using the conventional proteinase K digestion and phenol–chloroform extraction method.

Reagents:

1. Lysis buffer (100 ml) (50 mM Tris HCl, 50 mM EDTA, 100 mM NaCl, 1% SDS)—add 5 ml 50 mM Tris–HCL (pH 8.0), 10 ml 50 mM EDTA (pH 8.0), 2 ml 100 mM NaCl, and 10 ml 1% SDS (*see* **Note 1**). Mix thoroughly using a magnetic stirrer. Make the final volume to 100 ml with Milli-Q water.
2. Proteinase K (20 mg/ml).
3. Phenol:chloroform:isoamyl alcohol (25:24:1).
4. Chloroform.
5. Isopropanol or ethanol (for DNA precipitation).
6. 70% ethanol (for washing).

2.2 DNA Dissolution

The extracted DNA pellet is air-dried and dissolved in an adequate volume of TE buffer (10 mM Tris–HCL, pH 8.0; 0.1 mM EDTA).

2.3 PCR

1. Prepare 10 mM stock of each dNTP (dATP, dTTP, dCTP, dGTP).
2. Sterile nuclease-free water.

3 Methods

3.1 Sample Preparation

1. Collect the samples (insects) randomly for the experiment for the presence of different bacterial endosymbionts such as *Portiera*, *Wolbachia*, *Rickettsia*, *Arsenophonus*, and *Cardinium*.
2. Preserve the samples immediately in acetone for further analysis.
3. Extract the DNA from insects using the conventional proteinase K digestion and phenol extraction method.

Table 1
Primers and PCR cycling conditions for the identification of bacterial endosymbionts [16]

Endosymbiont	Primer sequence	Product Size	References
Portiera	F-5′TGCAAGTCGCGGCATCAT3′ R-5′CCGCCTTCTGCGTTGGCAACT3′	1000 bp	[16]
Wolbachia	F-5′CGGGGGAAAATTTATTGCT3′ R-5′AGCTGTAATACAGAAAGGAAA3′	650 bp	[16]
Rickettsia	F-5′GCTCAGAACGAACGCTGG3′ R-5′GAAGGAAAGCATCTCTGC3′	800 bp	[17]
Arsenophonus	F-5′CGTTTGATGAATTCATAGTCAAA3′ R-5′ GGTCCTCCAGTTAGTGTTACCCAAC3′	630 bp	[16]
Cardinium	F-5′GCGGTGTAAAATGAGCTTG3′ R-5′ACCTCTTCTTTAACTCAAGCCT3′	440 bp	[18]]
Buchnera	F-5′GAATTCTAGGTGTAGCGGTGA3′ R-5′GCGATTCCGACTTCGTGGA3′	0.66Kb	[19]
Serratia	F-5′AGAGTTTGATCMTGGCTCAG3′ R-5′GCAATGTCTTATTAACACAT3′	0.48Kb	[19]
Hamiltonella	F-5′GCAATGTCTTATTAACACAT3′ R-5′AGCACAGTTTACTGAGTTCA3′	0.2Kb	[19]
Regiella	F-5′ATCGGGGAGTAGCTTGCTAC3′ R-5′CTAGAGATCGTCGCCTAGGTA3′	0.2Kb	[19]
Rickettsia	F-5′AGAGTTTGATCMTGGCTCAG3′ R-5′CATCCATCAGCGATAAATCTTTC3′	0.2Kb	[19]
Spiroplasma	F-5′AGAGTTTGATCMTGGCTCAG3′ R-5′TAGCCGTGGCTTTCTGGTAA3′	0.51Kb	[19]

4. Dissolve the DNA in an adequate volume of TE buffer (10 mM Tris–HCl [pH 8.0], 0.1 mM EDTA).
5. Use specific bacterial primers for 16S rRNA gene amplification for bacterial genes (Table 1) [16].

3.2 PCR Reaction [16]

1. Perform the PCR reaction containing dNTPS (2.5 mM), 2.5 mM $MgCl_2$ (2.5 μl), Taq polymerase (1 U), forward and reverse primers, and DNA templates (25–30 ng) and make the final volume of 25 μl with Milli-Q water (*see* **Note 2**).
2. Set the PCR program as follows: Initial denaturation at 94 °C for 30 annealing temperatures (ranges from 50–58 °C), viz., for *Portiera* 58 °C, *Wolbachia* 52 °C, *Arsenophonus* 55 °C, *Rickettsia* 55 °C, and *Cardinium* 50 °C for 30 s (*see* **Note 3**).

3. Carry out the extension at 72 °C for 40 s and the final extension for 5 min at the same temperature for 35 cycles.
4. Run the PCR product in 0.8% agarose gel along with markers.

4 Notes

1. Store at *room temperature* or 4 °C depending on the duration of use (SDS may precipitate at low temperatures; warm gently to dissolve).
2. Perform PCR reactions with positive and negative controls: For positive control use 16s RNA of bacterial endosymbionts and without 16s RNA serve as negative control.
3. PCR products of different bacteria exhibit different band lengths (viz., *Portiera* 1000 bp, *Wolbachia* 650 bp, *Arsenophonus* 630 bp, *Rickettsia* 800 bp, *Cardinium* 440 bp).

References

1. Oakeson KF, Gil R, Clayton AL, Dunn DM, Von Niederhausern AC (2014) Genome degeneration and adaptation in a nascent stage of symbiosis. Genome Biol Evol 6:76–93. https://doi.org/10.1093/gbe/evt210
2. Gosalbes MJ, Latorre A, Lamelas A, Moya A (2010) Genomics of intracellular symbionts in insects. Int J Med Microbiol 300:271–278. https://doi.org/10.1016/j.ijmm.2009.12.001
3. Buchner P (1965) Endosymbiosis of animals and plant microorganisms. Wiley, New York
4. Baumann P (2005) Biology of bacteriocyte-associated endosymbionts of plant sap-sucking insects. Ann Rev Microbiol 59:155–189. https://doi.org/10.1146/annurev.micro.59.030804.121041
5. Dale C, Moran NA (2006) Molecular interactions between bacterial symbionts and their hosts. Cell 126:453–465
6. Bing XL, Yang L, Zchori-Fein E, Wang XW, Liu SS (2013) Characterization of a newly discovered symbiont of the whitefly *Bemisia tabaci* (Hemiptera: Aleyrodidae). Appl Environ Microbiol 79:569–575. https://doi.org/10.1128/AEM.03030-12
7. Baumann P, Moran NA (1997) Non-cultivable microorganisms from symbiotic associations of insects and other hosts. Antonie Leeuwenhoek 72:38–48
8. Clark MA, Baumann L, Munson MA, Baumann P, Campbell BC, Duffus JE, Osborne LS, Moran N (1992) The eubacterial endosymbionts of whiteflies (*Homoptera: Aleyrodoidea*) constitute a lineage distinct from the endosymbionts of aphids and mealybugs. Curr Microbiol 25:119–123
9. Fukatsu T, Ishikawa H (1996) Phylogenetic position of yeast-like symbiont of *Hamiltonaphis styraci* (*Homoptera, Aphididae*) based on 18SrDNA sequence. Insect Biochem Mol Biol 26:383–388
10. Fukatsu T, Nikoh N (1998) Two intracellular symbiotic bacteria from the mulberry psyllid *Anomoneura mori* (Insecta, Homoptera). Appl Environ Microbiol 64:3599–3606
11. Munson MA, Baumann P, Moran NA (1992) Phylogenetic relationships of the endosymbionts of mealybugs (*Homoptera: Pseudococcidae*) based on 16S rDNA sequences. Mol Phylogenet Evol 1:26–30
12. Munson MA, Baumann P, Clark MA, Baumann L, Moran NA, Voeglin DJ, Campbell BC (1991) Evidence for the establishment of aphid eubacterium endosymbiosis in an ancestor of four aphid families. J Bacteriol 173:6321–6324
13. Noda H, Nakashima N, Koizumi M (1995) Phylogenetic position of yeast-like symbiotes of rice planthoppers based on partial 18S rDNA sequences. Insect Biochem Mol Biol 25:639–646
14. Spaulding AW, Von Dohlen CD (1998) Phylogenetic characterization and molecular evolution of bacterial endosymbionts in psyllids

(Hemiptera: Sternorrhyncha). Mol Biol Evol 15:1506–1513

15. Unterman BM, Baumann P, McLean DL (1989) Pea aphid symbiont relationships established by analysis of 16S rRNAs. J Bacteriol 171:2970–2974
16. Raina HS, Singh A, Popli S, Pandey N, Rajagopal R (2015) Infection of bacterial endosymbionts in insects: a comparative study of two techniques viz PCR and FISH for detection and localization of symbionts in whitefly, *Bemisia tabaci*. PLoS One 10(8):e0136159. https://doi.org/10.1371/journal.pone.0136159
17. Gottlieb Y, Ghanim M, Chiel E, Gerling D, Portnoy V, Steinberg S (2006) Identification and localization of a rickettsia sp. in *Bemisia tabaci (Homoptera:Aleyrodidae)*. Appl Environ Microbiol 72:3646–3652. https://doi.org/10.1128/AEM.72.5.3646-3652.2006
18. Weeks AR, Breeuwer JAJ (2003) A new bacterium from the *cytophaga flavobacterium-* bacteroides phylum that causes sex ratio distortion. In: Bourtzisn K, Miller T (eds) Insect symbiosis II. CRC Press, Boca Raton, pp 165–176
19. Tsuchida T, Koga R, Sakurai M, Fukatsu T (2006) Facultative bacterial endosymbionts of three aphid species, *Aphis craccivora*, *Megoura crassicauda* and *Acyrthosiphon pisum*, sympatrically found on the same host plants. Appl Entomol Zool 41(1):129–113

Chapter 28

Fluorescent In Situ Hybridization for the Detection of Bacterial Infection of Endosymbionts in Insects

Sapna Chandwani, Hetvi Naik, and Natarajan Amaresan

Abstract

Several insect species show interaction with a number of bacteria, and these bacterial endosymbionts provide physiological, evolutionary, nutritional, and environmental benefits to the host (insects). To understand these bacterial symbiont–host interactions, different approaches have been used. Fluorescence in situ hybridization (FISH) is one of the approaches which provides a clear visualization of the bacterial endosymbionts residing inside the host tissues or cells. Not only this, but the technique can also be used to validate the findings and help provide a comprehensive understanding relationship between the host and symbionts.

Key words Endosymbionts, FISH, Bacterial infections, Insects

1 Introduction

To study bacterial infections in insects, the fluorescence in situ hybridization (FISH) technique is considered to be the most powerful tool as it has the ability to localize and detect specific bacterial sequences residing within the host tissues. There are many bacteria associated with the insects that cannot be cultured by using the standardized laboratory protocol. To address such challenges, the use of the FISH technique enables the identification of such unculturable bacteria by making use of fluorescently labeled probes [1]. These probes bind to the specific nucleic acid sequences in order to detect the endosymbionts in the host tissue (insects). This technique enables the researchers to determine the exact location of bacteria within the host tissues, their functions, and their ecological niches. This technique has immense potential in microbial ecology as it can identify the microbes and provide insights into the microbial distribution within complex communities [2].

Multiple bacterial species reside in the insects including both pathogens and symbionts. It is hard to differentiate specific

N. Amaresan and Sapna Chandwani (eds.), *Endosymbionts*, Springer Protocols Handbooks,
https://doi.org/10.1007/978-1-0716-4771-4_28,

bacterial strains, but FISH makes use of the fluorescent-labeled probes specific to their genetic sequences and differentiates such closely related bacterial species in a very appropriate manner providing the precise identification of such bacterial strains [1]. It can monitor bacterial infections at various stages, i.e., provides information regarding bacterial growth, progression of the infection, and spread of the infection within the host with respect to time [2]. Moreover, if this technique is combined with another imaging method such as confocal microscopy, it can provide a three-dimensional visualization of bacterial infections within the insect tissues [2]. In this context, this book chapter deals with the detection of bacterial infection of endosymbionts in insects via the FISH technique.

2 Materials

1. Insects where endosymbionts or pathogens are located from which specific tissues can be isolated.
2. Needles, forceps, or scissors.
3. Fluorescent microscope/confocal microscope.
4. Glasswares such as Petri dishes, pipette, glass bottles, flasks, microscopic glass slides, and micropipette.
5. *Fixatives:* Ethanol-based, Carnoy's fixative (ethanol:chloroform:glacial acetic acid, 6:3:1), or 4% paraformaldehyde.
6. *Media/Enzymes/Detergents:* Mounting media, phosphate-buffered saline (PBS), lysozyme, Triton ×-100, buffer containing formamide.
7. *Stains:* Fluorescent dyes such as fluorescein isothiocyanate (FITC), cyanine dye (Cy3).
8. *Instruments:* Centrifuge, incubator, RT-PCR, microscope.

3 Methods

3.1 Specimen Collection, Fixation, and Permeabilization

1. Collect the insect specimens where endosymbionts or pathogens are residing and dissect them with the help of needles, forceps, or scissors in order to isolate specific tissues such as reproductive organs or gut.
2. Now fix the collected tissue samples by using fixatives such as 4% paraformaldehyde or ethanol-based fixative or Carnoy's fixative in order to penetrate the probe as well as to permeabilize the cell wall and membranes (*see* **Note 1**).
3. Wash the samples in PBS after fixation and, in order to allow the probe to target rRNA within the bacterial cells, treat the

samples with detergents or enzymes such as Triton ×-100 or lysozyme [3].

3.2 Probe Preparation and Hybridization

4. To target specific bacterial 16S or 23S rRNA, label the probes with fluorescent dyes such as Cy3 or FITC to ensure selectivity, i.e., it can target specific genus or species of the bacteria.
5. Optimize and validate the probes in such a way that they can hybridize only with the targeted bacterial species instead of host or other microbial RNAs (*see* **Note 2**).
6. To hybridize, incubate the samples with the fluorescently labeled probes at 37–48 °C for several hours (*see* **Note 3**).
7. After hybridization, wash the samples with a buffer such as formamide in order to remove unbound probes (*see* **Note 4**) [4].

3.3 Sample Mounting and Visualization

8. Now, clean the glass slide, and mount the samples onto the microscopic glass slides using any of the mounting media containing antifade agents (*see* **Note 5**).
9. To determine the exact location of the bacterial infections pertaining within the tissues, visualize the fluorescent signal with the help of a confocal or fluorescent microscope.
10. Confocal or fluorescence microscopy will provide three-dimensional high-resolution images of the bacterial infections occurring within the tissues [5].

3.4 Data Analysis and Interpretation

11. To quantify bacterial distribution within the tissues, measure the fluorescence intensity, perform the cell counting, and analyze the images by using software such as quantitative PCR (q-PCR) (*see* **Note 6**).

4 Notes

1. To maintain the RNA integrity and to preserve the structure of the cell, tissue samples are allowed to fix.
2. Validating the probes means to ensure that the probe will bind only to the target bacteria, not to the nontarget bacterial species.
3. Optimal conditions (temperature and time) of sample incubation with probes depend on the targeted bacteria and probe design.
4. To prevent background fluorescence and nonspecific binding, stringent washing conditions are a must.
5. To prevent fluorophore photobleaching, the use of antifade agents is a must.

6. FISH along with q-PCR can quantify the abundance of the bacteria more precisely and accurately.

References

1. Raina HS, Singh A, Popli S, Pandey N, Rajagopal R (2015) Infection of bacterial endosymbionts in insects: a comparative study of two techniques *viz* PCR and FISH for detection and localization of symbionts in whitefly, *Bemisia tabaci*. PLoS One 10:e0136159
2. Volpi EV, Bridger JM (2008) FISH glossary: an overview of the fluorescence *in situ* hybridization technique. BioTechniques 45:385–409
3. Kliot A, Ghanim M (2016) Fluorescent *in situ* hybridization for the localization of viruses, bacteria and other microorganisms in insect and plant tissues. Methods 98:74–81
4. Amann RI, Ludwig W, Schleifer KH (1995) Phylogenetic identification and *in situ* detection of individual microbial cells without cultivation. Microbiol Rev 59:143–169
5. Bottari B, Ercolini D, Gatti M, Neviani E (2006) Application of FISH technology for microbiological analysis: current state and prospects. Appl Microbiol Biotechnol 73:485–494

Index

N. Amaresan and Sapna Chandwani (eds.), *Endosymbionts*, Springer Protocols Handbooks,
https://doi.org/10.1007/978-1-0716-4771-4,

C

D

E

F

G

H

I

L

M

N

O

P

T

U

V

W

X

Y

Z

MIX
Papier aus verantwortungsvollen Quellen
Paper from responsible sources
FSC® C105338

If you have any concerns about our products,
you can contact us on
ProductSafety@springernature.com

In case Publisher is established outside the EU,
the EU authorized representative is:
Springer Nature Customer Service Center GmbH
Europaplatz 3, 69115 Heidelberg, Germany

Printed by Libri Plureos GmbH
in Hamburg, Germany